METHODS FOR
ENVIRONMENTAL
TRACE ANALYSIS

MW

Analytical Techniques in the Sciences (AnTS)

Series Editor: David J. Ando, Consultant, Dartford, Kent, UK

A series of open learning/distance learning books which covers all of the major analytical techniques and their application in the most important areas of physical, life and materials science.

Titles Available in the Series

Analytical Instrumentation: Performance Characteristics and Quality
Graham Currell, University of the West of England, Bristol, UK

Fundamentals of Electroanalytical Chemistry
Paul M.S. Monk, Manchester Metropolitan University, Manchester, UK

Introduction to Environmental Analysis
Roger N. Reeve, University of Sunderland, UK

Polymer Analysis
Barbara H. Stuart, University of Technology, Sydney, Australia

Chemical Sensors and Biosensors
Brian R. Eggins, University of Ulster at Jordanstown, Northern Ireland, UK

Methods for Environmental Trace Analysis
John R. Dean, Northumbria University, Newcastle, UK

Forthcoming Titles

Analysis of Controlled Substances
Michael D. Cole, Anglia Polytechnic University, Cambridge, UK

Liquid Chromatography–Mass Spectrometry: An Introduction
Robert E. Ardrey, University of Huddersfield, Huddersfield, UK

METHODS FOR ENVIRONMENTAL TRACE ANALYSIS

John R. Dean
Northumbria University, Newcastle, UK

WILEY

Other Wiley Editorial Offices

John Wiley & Sons Inc., 111 River Street, Hoboken, NJ 07030, USA

Jossey-Bass, 989 Market Street, San Francisco, CA 94103-1741, USA

Wiley-VCH Verlag GmbH, Boschstr. 12, D-69469 Weinheim, Germany

John Wiley & Sons Australia Ltd, 33 Park Road, Milton, Queensland 4064, Australia

John Wiley & Sons (Asia) Pte Ltd, 2 Clementi Loop #02-01, Jin Xing Distripark, Singapore 129809

John Wiley & Sons Canada Ltd, 22 Worcester Road, Etobicoke, Ontario, Canada M9W 1L1

Wiley also publishes its books in a variety of electronic formats. Some content that appears
in print may not be available in electronic books.

Library of Congress Cataloging-in-Publication Data

Dean, John R.
    Methods for environmental trace analysis / John R. Dean.
        p. cm. – (Analytical techniques in the sciences)
    Includes bibliographical references and index.
    ISBN 0-470-84421-3 (cloth : alk. paper) – ISBN 0-470-84422-1 (pbk. : alk. paper)
        1. Pollutants – Analysis. 2. Trace analysis – Methodology. 3. Environmental
    chemistry – Methodology. 4. Sampling. I. Title. II. Series.

    TD193 .D43 2003
    628.5′028′7 – dc21
                                                                        2003028083

British Library Cataloguing in Publication Data

A catalogue record for this book is available from the British Library

ISBN 0-470-84421-3 (Cloth)
ISBN 0-470-84422-1 (Paper)

To Lynne, Sam and Naomi

Contents

Series Preface

There has been a rapid expansion in the provision of further education in recent years, which has brought with it the need to provide more flexible methods of teaching in order to satisfy the requirements of an increasingly more diverse type of student. In this respect, the *open learning* approach has proved to be a valuable and effective teaching method, in particular for those students who for a variety of reasons cannot pursue full-time traditional courses. As a result, John Wiley & Sons Ltd first published the Analytical Chemistry by Open Learning (ACOL) series of textbooks in the late 1980s. This series, which covers all of the major analytical techniques, rapidly established itself as a valuable teaching resource, providing a convenient and flexible means of studying for those people who, on account of their individual circumstances, were not able to take advantage of more conventional methods of education in this particular subject area.

Following upon the success of the ACOL series, which by its very name is predominately concerned with Analytical *Chemistry*, the *Analytical Techniques in the Sciences* (AnTS) series of open learning texts has now been introduced with the aim of providing a broader coverage of the many areas of science in which analytical techniques and methods are now increasingly applied. With this in mind, the AnTS series of texts seeks to provide a range of books which will cover not only the actual techniques themselves, but *also* those scientific disciplines which have a necessary requirement for analytical characterization methods.

Analytical instrumentation continues to increase in sophistication, and as a consequence, the range of materials that can now be almost routinely analysed has increased accordingly. Books in this series which are concerned with the *techniques* themselves will reflect such advances in analytical instrumentation, while at the same time providing full and detailed discussions of the fundamental concepts and theories of the particular analytical method being considered. Such books will cover a variety of techniques, including general instrumental analysis, spectroscopy, chromatography, electrophoresis, tandem techniques, electroanalytical methods, X-ray analysis and other significant topics. In addition, books in the series will include the *application* of analytical techniques in areas such as environmental science, the life sciences, clinical analysis, food science, forensic analysis, pharmaceutical science, conservation and archaeology, polymer science and general solid-state materials science.

Written by experts in their own particular fields, the books are presented in an easy-to-read, user-friendly style, with each chapter including both learning objectives and summaries of the subject matter being covered. The progress of the reader can be assessed by the use of frequent self-assessment question (SAQs) and discussion questions (DQs), along with their corresponding reinforcing or remedial responses, which appear regularly throughout the texts. The books are thus eminently suitable both for self-study applications and for forming the basis of industrial company in-house training schemes. Each text also contains a large amount of supplementary material, including bibliographies, lists of acronyms and abbreviations, and tables of SI Units and important physical constants, plus where appropriate, glossaries and references to literature sources.

It is therefore hoped that this present series of textbooks will prove to be a useful and valuable source of teaching material, both for individual students and for teachers of science courses.

Dave Ando
Dartford, UK

Preface

The field of environmental sample preparation has undergone a revolution in the last twenty five years. What was essentially a series of basic methods and procedures has developed (and continues to develop) into a new exciting area with a strong influence from instrumental approaches. This book essentially covers the traditional approaches of environmental sample preparation for both metals and organic compounds from a range of matrices.

The text is arranged into twelve chapters, covering the essentials of good laboratory housekeeping, through sampling and sample storage, and finally to the relevant sample preparation for inorganic and organic compounds from environmental matrices. A further chapter is devoted to the methods of analysis that can be used for quantitative analysis. To allow the user of the book to perform experiments in an effective manner, guidelines are also offered with respect to record keeping in the laboratory.

In Chapter 1, information is provided with regard to general safety aspects in the laboratory. In addition, specific guidance on the recording of numerical data (with the appropriate units) is provided, with examples on how to display data effectively in the form of tables and figures. Issues relating to sample handling of solids and liquids are also covered. Finally, numerical exercises involving the calculation of dilution factors and their use in calculating original concentrations in environmental samples are provided as worked examples.

Chapter 2 is concerned with the concept of quality assurance and all that it involves with respect to obtaining reliable data from environmental samples. Particular emphasis is placed on the definitions of accuracy and precision. Finally, details on the use of certified reference materials in environmental analysis are provided.

Chapter 3 involves the concept of sampling of representative sample systems. Specific details pertaining to the sampling of soil and sediment, water and air are provided. Chapter 4 considers the issues associated with the storage and preservation of samples with respect to inorganic and organic pollutants. In particular, focus is given to the retention of chemical species information in environmental matrices.

Chapters 5 and 6 are focused on the specific sample preparation approaches available for the elemental analysis of pollutants from environmental matrices, principally soil and water. Chapter 5 is concerned with the methods available to convert a solid environmental sample into the appropriate form for elemental analysis. The most popular methods are based on the acid digestion of the solid matrix, using either a microwave oven or a hot-plate approach. The growing importance of chemical species information is highlighted with some specific examples. This is then followed by examples of methods to selectively remove the species without destroying its speciation. Details are provided on the methods available for the selective extraction of metal species in soil studies using either a single extraction or a sequential extraction procedure. In addition, a procedure to carry out a physiologically based extraction test on soil is provided. Finally, the role of a simulated gastro-intestinal extraction procedure for extraction of metals in foodstuffs is provided. Chapter 6 provides details of methods for the extraction of metal ions from aqueous samples. Particular emphasis is placed on liquid–liquid extraction, with reference to ion-exchange and co-precipitation.

The focus in Chapters 7 and 8 is on the specific sample preparation approaches available for the extraction of organic compounds from environmental matrices, principally soil and water. Chapter 7 is concerned with the role of Soxhlet, ultrasonic and shake-flask extraction on the removal of organic compounds from solid (soil) matrices. These techniques are contrasted with newer developments in sample preparation for organic compound extraction, namely supercritical fluid extraction, microwave-assisted extraction and pressurized fluid extraction. Chapter 8 is arranged in a similar manner. Initially, details are provided on the use of solvent extraction for organic compounds removal from aqueous samples. This is followed by descriptions of the newer approaches, namely solid-phase extraction and solid-phase microextraction.

Chapter 9 deals with the extraction of volatile compounds from the atmosphere. Particular emphasis is placed here on the methods of thermal desorption and purge-and-trap. Chapter 10 focuses on the methods used to pre-concentrate samples after extraction. In this situation, particular attention is paid to two common approaches, namely rotary evaporation and gas 'blow-down', although details of two other methods are also provided.

Chapter 11 details the relevant methods of analysis for both metals and organic compounds. For elemental (metal) analysis, particular attention is given to atomic spectroscopic methods, including atomic absorption and atomic emission spectroscopy. Details are also provided on X-ray fluorescence spectrometry for the direct analysis of metals in solids, ion chromatography for anions in solution, and anodic stripping voltammetry for metal ions in solution. For organic compounds, particular attention is focused on chromatographic approaches, principally gas chromatography and high performance liquid chromatography. Details are also provided on the use of Fourier-transform infrared spectroscopy for the analysis of total petroleum hydrocarbons.

The final chapter (Chapter 12) provides examples of forms that could be used to record laboratory information at the time of doing the experiment. Guidelines are given for the recording of information associated with sample pre-treatment. Then, specific forms are provided for the recording of sample preparation details associated with inorganic or organic environmental samples. Finally, guidelines are given for the recording of information associated with the analysis of metals and organic compounds. This chapter concludes with a resource section detailing lists of journals, books (general and specific), CD-ROMs, videos and Web addresses that will act to supplement this text.

Finally, I should like to give a special mention to all of the students (both past and present) who have contributed to the development of interest in the field of environmental sample preparation. The achievements have been many and varied across a broad area of environmental sample preparation, but it has all been worthwhile.

John R. Dean
Northumbria University, Newcastle, UK

Acronyms, Abbreviations and Symbols

AAS	atomic absorption spectroscopy
AC	alternating current
ACN	acetonitrile
ACS	American Chemical Society
A/D	analogue-to-digital (converter)
AE	alcohol ethoxylate
AES	atomic emission spectroscopy
AFS	atomic fluorescence spectroscopy
aMAE	atmospheric microwave-assisted extraction
ANOVA	analysis of variance
APDC	ammonium pyrrolidine dithiocarbamate
AsB	arsenobetaine
AsC	arsenocholine
ASE™	accelerated solvent extraction
ASV	anodic stripping voltammetry
atm	atmosphere (unit of pressure)
BCR	(European) Community Bureau of Reference
BOD	biochemical oxygen demand
BPR	back-pressure regulator (restrictor)
bpt	boiling point
BTEX	benzene−toluene−ethylbenzene−xylene(s) (mixture)
C	coulomb
CCD	charge-coupled device
CE	capillary electrophoresis
CEC	capillary electrochromatography
cGC	capillary gas chromatography

CI	chemical ionization
COD	chemical oxygen demand
COSHH	Control of Substances Hazardous to Health
CRM	Certified Reference Material
CVAAS	cold-vapour atomic absorption spectroscopy
ECD	electron-capture detector (detection)
EDTA	ethylenediaminetetraacetic acid
EI	electron impact
emf	electromotive force
ETAAS	electrothermal (atomization) atomic absorption spectroscopy
eV	electronvolt
EVACS	(automated) evaporative concentration system
FAAS	flame atomic absorption spectroscopy
FID	flame ionization detector (detection)
FL	fluorescence (detection)
FP	flame photometry
FPD	flame photometric detector (detection)
FTIR	Fourier-transform infrared (spectroscopy)
GC	gas chromatography
GFAAS	graphite-furnace atomic absorption spectroscopy
HASAW	Health and Safety at Work (Act)
HCL	hollow-cathode lamp
HPLC	high performance liquid chromatography
HyAAS	hydride-generation atomic absorption spectroscopy
IC	ion chromatography
ICP	inductively coupled plasma
ICP–AES	inductively coupled plasma–atomic emission spectroscopy
ICP–MS	inductively coupled plasma–mass spectrometry
id	internal diameter
IR	infrared
ISO	International Organization for Standardization
IUPAC	International Union of Pure and Applied Chemistry
J	joule
LC	liquid chromatography
LDR	linear dynamic range
LGC	Laboratory of the Government Chemist
LLE	liquid–liquid extraction
MAE	microwave-assisted extraction
MBT	monobutyltin
MIBK	methylisobutyl ketone (4-methylpentan-2-one)
MIP	microwave-induced plasma
MMAA	monomethylarsonic acid
MS	mass spectrometry

MSD	mass-selective detector
MSPD	matrix solid-phase dispersion
NIST	National Institute of Standards and Technology
NMR	nuclear magnetic resonance (spectroscopy)
OCP	organochlorine pesticide
ODS	octadecylsilane
OPP	organophosphate pesticide
OT(s)	organotin(s)
PAH	polycyclic (polynuclear) aromatic hydrocarbon
PBET	physiologically based extraction test
PCB	polychlorinated biphenyl
PCDD	polychlorinated dibenzo-p-dioxin
PCDF	polychlorinated dibenzofuran
PFE	pressurized fluid extraction
pMAE	pressurized microwave-assisted extraction
PMT	photomultiplier tube
ppb	parts per billion (10^9)
ppm	parts per million (10^6)
ppt	parts per thousand (10^3)
psi	pounds per square inch
PTFE	polytetrafluoroethylene
QFAAS	quartz-furnace atomic absorption spectroscopy
RF	radiofrequency
rms	root mean square
rpm	revolutions per minute
RSC	The Royal Society of Chemistry
SAX	strong anion exchange
SCX	strong cation exchange
SD	standard deviation
SE	standard error
SFC	supercritical fluid chromatography
SFE	supercritical fluid extraction
SI (units)	Système International (d'Unitès) (International System of Units)
SIM	single-ion monitoring
SM&T	Standards, Materials and Testing
SOM	soil organic matter
SPDC	sodium pyrrolidinedithiocarbamate
SPE	solid-phase extraction
SPME	solid-phase microextraction
SRM	Standard Reference Material
TBT	tributyltin
TBTO	tributyltin oxide
TCD	thermal conductivity detector

TGA	thermal gravimetric (thermogravimetric) analysis
TLC	thin layer chromatography
TPH	total petroleum hydrocarbon
TPT	triphenyltin
URL	uniform resource locator
USEPA	United States Environmental Protection Agency
UV	ultraviolet
V	volt
vis	visible
VOA	volatile organic analyte
W	watt
WWW	World Wide Web
XRF	X-ray fluorescence (spectroscopy)

A_r	relative atomic mass
C	speed of light; concentration
D	distribution ratio
e	electronic charge
E	energy; electric-field strength; fraction of analyte extracted
E_h	redox potential
f	(linear) frequency; focal length
F	Faraday constant
G	gravitational constant
I	electric current
K	partition coefficient
K_d	distribution coefficient
m	mass
M_r	relative molecular mass
p	pressure
Q	electric charge (quantity of electricity)
R	molar gas constant; resistance; correlation coefficient
R^2	coefficient of determination
t	time; Student factor; statistical (theoretical) significance
T	thermodynamic temperature
V	electric potential
z	ionic charge

λ	wavelength
ν	frequency (of radiation)
σ	measure of standard deviation
σ^2	variance

About the Author

John R. Dean, B.Sc., M.Sc., Ph.D., D.I.C., D.Sc., FRSC, CChem, Registered Analytical Chemist

John R. Dean took his first degree in Chemistry at the University of Manchester Institute of Science and Technology (UMIST), followed by an M.Sc. in Analytical Chemistry and Instrumentation at Loughborough University of Technology, and finally a Ph.D. and D.I.C. in Physical Chemistry at Imperial College of Science and Technology (University of London). He then spent two years as a postdoctoral research fellow at the Food Science Laboratory of the Ministry of Agriculture, Fisheries and Food in Norwich, in conjunction with The Polytechnic of the South West in Plymouth. His work there was focused on the development of directly coupled high performance liquid chromatography and inductively coupled plasma–mass spectrometry methods for trace element speciation in foodstuffs. This was followed by a temporary lectureship in Inorganic Chemistry at Huddersfield Polytechnic. In 1988, he was appointed to a lectureship in Inorganic/Analytical Chemistry at Newcastle Polytechnic (now Northumbria University). This was followed by promotion to Senior Lecturer (1990), Reader (1994) and Principal Lecturer (1998). In 1995 he was the recipient of the 23rd Society for Analytical Chemistry (SAC) Silver Medal, and was awarded a D.Sc. (University of London) in Analytical and Environmental Science in 1998. He has published extensively in analytical and environmental science. He is an active member of the Royal Society of Chemistry (RSC) Analytical Division, having served as a member of the Atomic Spectroscopy Group for 15 years (10 as Honorary Secretary), as well as a past Chairman (1997–1999). He has served on the Analytical Division Council for three terms and is currently its Vice-President (2002–2004), as well as the present Chairman of the North-East Region of the RSC (2001–2003).

Chapter 1
Basic Laboratory Skills

Learning Objectives

- To be aware of safety aspects in the laboratory.
- To be able to record, in an appropriate style, practical information accurately.
- To be able to record numerical data with appropriate units.
- To understand the importance of sample handling with respect to both solids and liquids.
- To be able to present data effectively in tables and figures.
- To be able to perform numerical exercises involving dilution factors.

1.1 Introduction

All scientific studies involve some aspect of practical work. It is therefore essential to be able to observe and to record information accurately. In the context of environmental analyses, it should be borne in mind that not all practical work is carried out in the laboratory. Indeed it could be argued that the most important aspects of the whole practical programme are done outside the laboratory in the field, as this is the place where the actual sampling of environmental matrices (air, water, soil, etc.) takes place. It is still common practice, however, to transport the acquired sample back to the laboratory for analysis, so knowledge and implementation of the storage conditions and containers to be used are important. Both sampling and sample storage are covered in Chapters 3 and 4, respectively.

1.2 Safety Aspects

No laboratory work should be carried out without due regard to safety, both for yourself and for the people around you. While the Health and Safety at Work

Act (1974) provides the main framework for health and safety, it is the Control of Substances Hazardous to Health (COSHH) regulations of 1994 and 1996 that impose strict legal requirements for risk assessment wherever chemicals are used. Within this context, the use of the terms *hazard* and *risk* are very important. A hazardous substance is one that has the ability to cause harm, whereas risk is about the likelihood that the substance may cause harm. Risk is often associated with the quantity of material being used. For example, a large volume of a flammable substance obviously poses a greater risk than a very small quantity. Your laboratory will operate its own safety scheme, so ensure that you are aware of what it is and follow it.

The basic rules for laboratory work (and, as appropriate, for associated work outside the laboratory using chemicals) are as follows:

- Always wear appropriate protective clothing. Typically, this involves a clean laboratory coat fastened up, eye protection in the form of safety glasses or goggles, appropriate footwear (open-toed sandals or similar are inappropriate) and ensure that long hair is tied back. In some circumstances, it may be necessary to put on gloves, e.g. when using strong acids.
- Never smoke, eat or drink in the laboratory.
- Never work alone in a laboratory.
- Make yourself familiar with the fire regulations in your laboratory and building.
- Be aware of the accident/emergency procedures in your laboratory and building.
- Never mouth pipettes – use appropriate devices for transferring liquids.
- Only use/take the minimum quantity of chemical required for your work.
- Use a fume cupboard for hazardous chemicals. Check that it is functioning properly before starting your work.
- Clear up spillages on and around equipment and in your own workspace as they occur.
- Work in a logical manner.
- Think ahead and plan your work accordingly.

DQ 1.1

What is one of the first things that you should consider before starting a laboratory experiment?

Answer

You should make yourself aware of the particular safety aspects that operate in your own laboratory. This includes the position of fire safety equipment, the methods of hazard and risk assessments for the chemicals to be used, the use of fume cupboards, fire regulations and evacuation procedures, and the disposal arrangements for used chemicals.

1.3 Recording of Practical Results

This is often done in an A4 loose-leaf binder, which offers the flexibility to insert graph paper at appropriate points. Such binders do, however, have one major drawback in that pages can be lost. Bound books obviously avoid this problem. All experimental observations and data should be recorded in the notebook – in ink – at the same time that they are made. It is easy to forget information when you are busy!

The key factors to remember are as follows:

- Record data correctly and legibly.
- Include the date and title of individual experiments.
- Outline the purpose of the experiment.
- Identify and record the hazards and risks associated with the chemicals/equipment being used.
- Refer to the method/procedure being used and/or write a brief description of the method.
- Record the *actual* observations and not your *own* interpretation, e.g. the colour of a particular chemical test – unfortunately, colour can be subjective. In this situation, it is possible to use the *Munsell Book of Colour*. This is a master atlas of colour that contains almost 1600 colour comparison chips. The colours are prepared according to an international standard. There are 40 pages, with each being 2.5 hue steps apart. On each page, the colour chips are arranged by Munsell value and chroma. The standard way to describe a colour using Munsell notation is to write the numeric designation for the Munsell hue (H) and the numeric designation for value (V) and chroma (C) in the form H V/C.
- Record numbers with the correct units, e.g. mg, g, etc., and to an appropriate number of significant figures.
- Interpret data in the form of graphs, spectra, etc.
- Record conclusions.
- Identify any actions for future work.

1.4 Units

The Système International d'Unitès (SI) is the internationally recognized system for measurement. This essentially uses a series of base units (Table 1.1) from which other terms are derived. The most commonly used SI derived units are shown in Table 1.2. It is also common practice to use prefixes (Table 1.3) to denote multiples of 10^3. This allows numbers to be kept between 0.1 and 1000. For example, 1000 ppm (parts per million) can also be expressed as 1000 $\mu g\,ml^{-1}$, 1000 $mg\,l^{-1}$ or 1000 $ng\,\mu l^{-1}$.

Table 1.1 The base SI units

Measured quantity	Name of unit	Symbol
Length	Metre	m
Mass	Kilogram	kg
Amount of substance	Mole	mol
Time	Second	s
Electric current	Ampere	A
Thermodynamic Temperature	Kelvin	K
Luminous intensity	Candela	cd

Table 1.2 SI derived units

Measured quantity	Name of unit	Symbol	Definition in base units	Alternative in derived units
Energy	Joule	J	$m^2\,kg\,s^{-2}$	$N\,m$
Force	Newton	N	$m\,kg\,s^{-2}$	$J\,m^{-1}$
Pressure	Pascal	Pa	$kg\,m^{-1}\,s^{-2}$	$N\,m^{-2}$
Electric charge	Coulomb	C	$A\,s$	$J\,V^{-1}$
Electric potential difference	Volt	V	$m^2\,kg\,A^{-1}\,s^{-3}$	$J\,C^{-1}$
Frequency	Hertz	Hz	s^{-1}	—
Radioactivity	Becquerel	Bq	s^{-1}	—

Table 1.3 Commonly used prefixes

Multiple	Prefix	Symbol
10^{18}	exa	E
10^{15}	peta	P
10^{12}	tera	T
10^9	giga	G
10^6	mega	M
10^3	kilo	k
10^{-3}	milli	m
10^{-6}	micro	μ
10^{-9}	nano	n
10^{-12}	pico	p
10^{-15}	femto	f
10^{-18}	atto	a

SAQ 1.1

The prefixes shown in Table 1.3 are frequently used in environmental science to represent large or small quantities. Convert the following quantities by using the suggested prefixes.

Quantity	m	μm	nm
6×10^{-7} m			
Quantity	mol l^{-1}	mmol l^{-1}	μmol l^{-1}
2.5×10^{-3} mol l^{-1}			
Quantity	μg ml^{-1}	mg l^{-1}	ng μl^{-1}
8.75 ppm			

1.5 Sample Handling: Liquids

The main vessels used for measuring out liquids in environmental analyses can be sub-divided into those used for quantitative work and those used for qualitative work. For the former, we frequently use volumetric flasks, burettes, pipettes and syringes, and for the latter, beakers, conical flasks, measuring cylinders, test tubes and Pasteur pipettes.

The nature of the vessel may be important in some instances. For example, some plasticizers are known to leach from plastic vessels, especially in the presence of organic solvents, e.g. dichloromethane. This is particularly important in organic analyses. In inorganic analyses, contamination risk is evident from glass vessels that may not have been cleaned effectively. For example, metal ions can adsorb to glass and then leach into solution under acidic conditions, thereby causing contamination. This can be remedied by cleaning the glassware prior to use by soaking for 24 h in 10% nitric acid solution, followed by rinsing with deionized water (three times). The cleaned vessels should then either be stored upside down or covered with Clingfilm® to prevent dust contamination.

1.6 Sample Handling: Solids

The main vessels used for weighing out solids in environmental analyses are weighing bottles, plastic weighing dishes or weighing boats. These containers are used to accurately weigh the solid, using a four-decimal-place balance, and to transfer a soluble solid directly into a volumetric flask. If the solid is not totally soluble it is advisable to transfer the solid to a beaker, add a suitable solvent, e.g. deionized or distilled water, and stir with a clean glass rod until all of the solid has dissolved. It may be necessary to heat the solution to achieve complete dissolution. Then quantitatively transfer the cooled solution to the volumetric flask and make up to the graduation mark with solvent. NOTE – volumetric flasks are accurate for their specified volume when the solution itself is at a particular temperature, e.g. 20°C.

1.7 Preparing Solutions for Quantitative Work

Solutions are usually prepared in terms of their molar concentrations, e.g. $mol\,l^{-1}$, or mass concentrations, e.g. $g\,l^{-1}$. It should be noted that both of these refer to an amount per unit volume, i.e. concentration = amount/volume. It is important to use the highest (purity) grade of chemicals (liquids or solids) for the preparation of solutions for quantitative analysis, e.g. AnalaR® or AristaR®. For example, consider the preparation of a 1000 ppm solution of lead from its metal salt.

NOTE: the molecular weight of $Pb(NO_3)_2 = 331.20$; the atomic weight of $Pb = 207.19$.

$$\frac{331.20}{207.19} = 1.5985 \ g \ of \ Pb(NO_3)_2 \ in \ 1 \ litre$$

Therefore, dissolve 1.5985 g of $Pb(NO_3)_2$ in 1 vol% HNO_3 (AnalaR® or equivalent) and dilute to one litre in 1 vol% HNO_3. This will give you a 1000 ppm solution of Pb.

1.8 Presentation of Data: Tables

A useful method of recording numerical data is in the form of a table. All tables should have a title that adequately describes the data presented (they may need to be numbered so that they can be quoted in the text). It is important to display the components of the table such that it allows direct comparison of data and to allow the reader to easily understand the significance of the results. It is normal to tabulate data in the form of columns and rows, with columns running vertically and rows horizontally. Columns contain, for example, details of concentration and units, sampling sites or properties measured, while rows contain numerical or written descriptions for the columns. The first column often contains the independent variable data, e.g. concentration or site location, while subsequent columns may contain numerical values of concentrations for different metals or organics. A typical tabulated set of data obtained from an experiment to determine the level of lead in soil by using atomic absorption spectroscopy is shown in Table 1.4.

Table 1.4 Calibration data obtained for the determination of lead in soil by using atomic absorption spectroscopy

Concentration (ppm)	Absorbance
0	0.000
2	0.015
4	0.032
6	0.045
8	0.062
10	0.075

It is important when tabulating or graphing (see below) data to not quote values to more significant figures than is necessary.

1.9 Presentation of Data: Graphs

The common usage of computers means that graphs are now most frequently produced by using computer-based graphics packages. However, irrespective of the mode of preparation, it is important to ensure that the graph is correctly presented. All graphs should have a title that adequately describes the data presented (they may need to be numbered so that they can be quoted in the text). Most graphs are used to describe a relationship between two variables, e.g. x and y. It is normal practice to identify the x-axis as the horizontal (abscissa) axis and to use this for the independent variable, e.g. concentration. The vertical (or ordinate) axis (y-axis) is therefore used to plot the dependent variable, e.g. concentration response. Each axis should contain a descriptive label indicating what is being represented, together with the appropriate units of measurement.

The mathematical relationship most commonly used for calibration is of the following form:

$$y = mx + c$$

where y is the signal response, e.g. absorbance or signal (mV), x is the concentration of the working solution (in appropriate units, e.g. $\mu g\,ml^{-1}$ or ppm), m is the slope of the graph, and c is the intercept on the x-axis.

A typical graphical representation of the data obtained from an experiment to determine the level of lead in soil by using atomic absorption spectroscopy is shown in Figure 1.1 (also tabulated in Table 1.4 above).[†]

[†] R, in this figure (and also in Figure 1.2), is known as the *correlation coefficient*, and provides a measure of the quality of calibration. In fact, R^2 (*the coefficient of determination*) is used because it is more sensitive to changes. This varies between -1 and $+1$, with values very close to -1 and $+1$ pointing to a very tight 'fit' of the calibration curve.

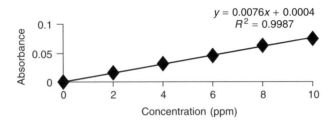

Figure 1.1 Determination of lead in soil: direct calibration approach.

SAQ 1.2

Graphically plot the following data, ensuring that you correctly identify and label the axes. These data were obtained for the analysis, by gas chromatography with flame ionization detection, of lindane in waste water. The signal for lindane has been ratioed to the internal standard. This acts to compensate for any variations in injection volume that are inherent in injecting 1 μl of sample solution.

Concentration (ppm)	Lindane/internal standard
0	0.00
2	0.39
4	0.71
6	1.15
8	1.54
10	1.84

An alternative approach to undergoing a *direct calibration*, as has just been described, is the use of the *method of standard additions*. This may be particularly useful if the sample is known to contain a significant portion of a potentially interfering matrix. In standard additions, the calibration plot no longer passes through zero (on both the x- and y-axes). As the concept of standard additions is to eliminate any matrix effects present in the sample, it is not surprising to find that the working standard solutions all now contain the same volume of the sample, i.e. the same volume of the sample is introduced into a succession of working solutions. Each of the working solutions, containing the same volume of the sample, is then introduced into the instrument and the response is again recorded as before. However, 'graphing' the signal response (e.g. absorbance, signal (mV), etc.) against analyte concentration in this case produces a very different type of plot. In this situation, the graph no longer passes through zero on either axis, but if correctly drawn, the graph can be extended towards the x-axis (extrapolated) until it intercepts it. By maintaining a constant concentration x-axis, the unknown sample concentration can be determined (Figure 1.2). It is

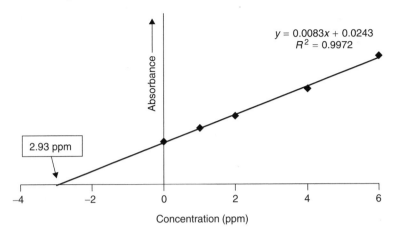

Figure 1.2 Determination of lead in soil: standard additions method.

essential that this graph is linear over its entire length or otherwise considerable errors can be introduced.

DQ 1.2

You have successfully determined the levels of lindane in waste water using gas chromatography. Should you graph the data or record the data in the form of a table?

Answer

The simple answer is that you are trying to convey to the reader information on what you have observed experimentally in the most effective manner. You may decide that the most effective means to convey your message is to plot a graph; this is particularly useful if you have a lot of numerical data based on, for example, responses to different concentrations. On the other hand, if you have more than two variables (remember you are plotting a graph in two dimensions only) it is probably easiest to tabulate the data, for example, responses to different concentrations at different sampling sites. As you can probably guess, even the latter can be represented in graphical form by simply plotting more then one trend line on one graph using different colours (for clarity).

1.10 Calculations: Dilution Factors

1. Calculate the concentration (in μg g^{-1}) of *lead* in the soil sample described above. An accurately weighed (2.1270 g) soil sample is digested in 25 ml of concentrated nitric acid, cooled and then quantitatively transferred to a 100 ml

volumetric flask and made up to the mark with distilled water. This solution is then diluted by taking 10 ml of the solution and transferring to a further 100 ml volumetric flask where it is made up to the mark with high-purity water. What is the dilution factor?

$$\frac{100 \ ml}{2.1270 \ g} \times \frac{100 \ ml}{10 \ ml} = 470 \ ml \ g^{-1}$$

If the solution was then analysed and found to be within the linear portion of the graph (see Figure 1.1), the value for the dilution factor would then be multiplied by the concentration from the graph, thus producing a final value representative of the element under investigation.

SAQ 1.3

What is the concentration of lead in the original soil sample? If the absorbance from the digested sample was 0.026, calculate the concentration of lead from the graph and then apply the dilution factor.

---

2. Calculate the concentration (in $\mu g \ ml^{-1}$) of lindane in the waste water sample discussed in SAQ 1.2 above. A waste water sample (1000 ml) was extracted into dichloromethane (3 × 10 ml) using liquid–liquid extraction. The extract was then quantitatively transferred to a 50 ml volumetric flask and made up to the mark with dichloromethane. What is the dilution factor?

$$\frac{50 \ ml}{1000 \ ml} = 0.05 \ ml \ ml^{-1}$$

If the solution was then analysed and found to be within the linear portion of the graph (see SAQ 1.2), the value for the dilution factor would then be multiplied by the concentration from the graph, thus producing a final value representative of the element under investigation.

SAQ 1.4

What is the concentration of lindane in the waste water sample? If the ratio of lindane to internal standard from the extracted sample was 0.26, calculate the concentration of lindane from the graph and then apply the dilution factor.

Summary

A good set of practical notes should provide the following:

- A brief indication of what you hope to achieve by carrying out the work, i.e. the aims of the experiment.

- A record of all of the chemicals/reagents (and their purities/grades) used in the work.
- A record of all of the equipment/apparatus used in the work and their experimental settings. For example, when using flame atomic absorption spectroscopy it is important to record the make and model of instrument used, the metal to be determined, the wavelength used, and the flame constituents and their flow rates. If using gas chromatography, for example, it is important to identify the make and model of instrument being used, the type of detector being employed, the column and its dimensions, the carrier gas and its flow rate, and the retention time(s) of your peak(s) of interest.

In addition, it is important to:

- Record all data immediately. At this stage, a simple table can be used to accurately record the data.
- Note any immediate conclusions and possible suggestions for future work.

Further Reading

Dean, J. R., Jones, A. M., Holmes, D., Reed, R., Weyers, J. and Jones, A., *Practical Skills in Chemistry*, Prentice Hall, Harlow, UK, 2002.

Chapter 2

Investigative Approach for Sample Preparation

Learning Objectives

- To understand the concept of quality assurance and all that it involves with respect to obtaining good data.
- To understand the differences between accuracy and precision and be able to use them appropriately.
- To appreciate the concept of a certified reference material and be able to understand when one is required.

2.1 Introduction

The key to effective laboratory work is the planning and organization of the experimental work prior to commencement. A major factor associated with the planning process involves the preparation of a clean work environment and equipment. It is also important during the laboratory work to conform to good housekeeping, i.e. clean and store equipment appropriately after use so that it is ready for future work. Contamination is often the unseen barrier for all environmental analyses and can lead to false results being obtained.

In addition, it is important to consider the following:

- Store equipment/apparatus appropriately, e.g. beakers should be stored either open end down on a clean surface or by covering their openings with an inert material, e.g. Clingfilm®.

- Glassware should never be left on benches and/or in fume cupboards.

- Label all cupboards and/or drawers where equipment/apparatus is to be stored.
- Always return chemicals to their correct storage location immediately after usage.
- Ensure chemicals are stored correctly, e.g. some chemicals are light sensitive – these must be stored in dark-coloured bottles; acids should be kept in safety cabinets. Always check the manufacturers/suppliers guidelines for chemical storage.
- Always correctly dispose of chemicals appropriately, e.g. organic solvents require special disposal arrangements – often arranged by the departmental safety officer.
- Consider other people who may be working nearby.

2.2 Quality Assurance

Quality assurance is about getting the *correct result*. In environmental analysis and monitoring, this involves several steps, including sample collection, treatment and storage, followed by laboratory analysis. A complete environmental protocol is shown in Figure 2.1. It is likely that the variation in the final measurement is more influenced by the work external to the analytical laboratory than that within the laboratory. Two important terms in quality assurance are *accuracy* and *precision*.

Accuracy is the closeness of a determined value to its 'true' value, while precision is the closeness of the determined values to each other. A determined result for the analysis of a polycyclic aromatic hydrocarbon in soil could produce precise (i.e. repeatable) but inaccurate (i.e. untrue) results.

SAQ 2.1

If the black circles shown in Figure 2.2 represent the results obtained and the centre of the rings represents the 'true' values, which of the results are accurate and which are precise?

In order to achieve good accuracy and precision – in the laboratory at least – it is desirable that a good quality assurance scheme is operating. The main objectives of such a scheme are as follows:

- to select and validate appropriate methods of sample preparation
- to select and validate appropriate methods of analysis
- to maintain and upgrade analytical instruments
- to ensure good record keeping of methods and results
- to ensure the quality of data produced
- to maintain a high quality of laboratory performance

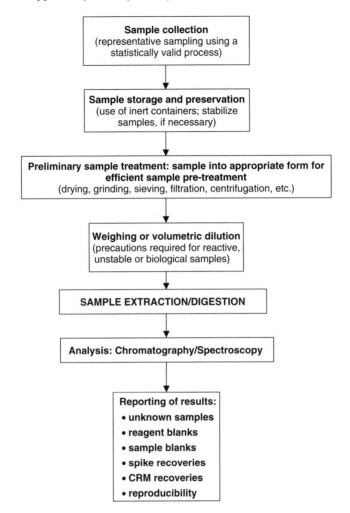

Figure 2.1 A typical analytical protocol used for environmental analysis.

In implementing a good quality control programme, it is necessary to take into account the following:

- *Certification of analyst competence*. This is intended to assess whether a particular analyst can carry out sample and standard manipulations, operate the instrument in question and obtain data of appropriate quality. The definition of suitable data quality is open to interpretation but may be assessed in terms of replicate analyses of a 'check sample'.
- *Recovery of known additions*. Samples are spiked with known concentrations of the same analyte and their recoveries noted. This approach will also allow

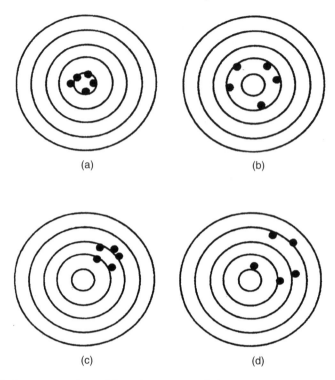

Figure 2.2 Accuracy and precision: the centres of the bull's-eyes represent the 'true' values (cf. SAQ 2.1).

the operator to determine whether any matrix effects are interfering with the analysis.

- *Analysis of certified reference materials.* By definition, a certified reference material (CRM) is a substance for which one or more analytes have certified values, produced by a technically valid procedure, accompanied with a traceable certificate and issued by a certifying body (e.g. Figure 2.3).

DQ 2.1

Consider the quoting of numerical values and units for the major/minor constituents and the trace constituents in the CRM of Figure 2.3.

Answer

It is noted that numerical values are often grouped so that the numbers are within 1000 of each other with the appropriate choice of units, e.g. 0.013 $\mu g\, g^{-1}$ cadmium is within a factor of 1000 for zinc (12.5 $\mu g\, g^{-1}$) but not within a factor of 1000 for phosphorus (0.159 wt% or 1590 $\mu g\, g^{-1}$).

National Institute of Science and Technology
Certificate of Analysis

Standard Reference Material 1515
Apple Leaves

Certified Concentrations of Constituent Elements[1]

Element	Concentration (wt%)
Calcium	1.526 ± 0.015
Magnesium	0.271 ± 0.008
Nitrogen (total)	2.25 ± 0.19
Phosphorus	0.159 ± 0.011
Potassium	1.61 ± 0.02

Element	Concentration (μg g^{-1})[2]	Element	Concentration (μg g^{-1})[2]
Aluminium	286 ± 9	Mercury	0.044 ± 0.004
Arsenic	0.038 ± 0.007	Molybdenum	0.094 ± 0.013
Barium	49 ± 2	Nickel	0.91 ± 0.12
Boron	27 ± 2	Rubidium	10.2 ± 1.5
Cadmium	0.013 ± 0.002	Selenium	0.050 ± 0.009
Chlorine	579 ± 23	Sodium	24.4 ± 12
Copper	5.64 ± 0.24	Strontium	25 ± 2
Iron	83 ± 5	Vanadium	0.26 ± 0.03
Lead	0.470 ± 0.024	Zinc	12.5 ± 0.3
Manganese	54 ± 3		

[1]The certified concentrations are equally weighted means of results from two or more different analytical methods or the means of results from a single method of known high accuracy.

[2]The values are based on dry weights. Samples of this SRM must be dried before weighing and analysis by, for example, drying in a desiccator at room temperature (ca. 22 °C) for 120 h over fresh anhydrous magnesium perchlorate. The sample depth should not exceed 1 cm.

Figure 2.3 An example of a certificate of analysis for elements in apple leaves. Reprinted from Certificate of Analysis, *Standard Reference Material 1515, Apple Leaves*, National Institute of Standards and Technology. Not copyrightable in the United States.

In addition, for the major/minor constituents three significant figures are quoted for values in the wt% range. This allows large concentrations to be quoted without the necessity to report excessive significant figures, e.g. instead of quoting 15 260 μg g^{-1} for calcium, it is more appropriate to quote 1.526 wt%.

> *All values are quoted with a variation (±) of one standard deviation of the mean value.*

Some common examples of certifying bodies are the National Institute for Standards and Technology (NIST) based in Washington, DC, USA, the Community Bureau of Reference (known as BCR), in Brussels, Belgium, and the Laboratory of the Government Chemist (LGC), in Teddington, Middlesex, U.K. CRMs can be either pure materials or matrix materials. Pure materials are used for the calibration of instruments, whereas matrix materials are used for the validation of a whole method from sample preparation through to the final measurement. A list of the commonly available matrix CRMs for environmental analyses is given in Table 2.1.

Table 2.1 A selection of the commonly available matrix certified reference materials for environmental analyses [1][a]

Code	Name	Description
Water		
LGC6010	Hard drinking water	Certified values for 16 metals in acidified (pH < 2) tap water (Teddington, Middlesex, UK)
LGC6011	Soft drinking water	Certified values for 13 metals in acidified (pH < 2) tap water (Merthyr Tydfil, Wales)
LGC6012	Hard drinking water	Certified values for 4 anions in acidified (pH < 2) tap water (Teddington, Middlesex, UK)
LGC6013	Soft drinking water	Certified values for 4 anions in acidified (pH < 2) tap water (Merthyr Tydfil, Wales)
LGC6017	Rainwater (run-off)	Certified values for 11 metals in acidified (pH 2) roof run-off rainwater (Kingston-upon-Thames, Surrey, UK)
LGC6018	Rainwater (run-off)	Certified values for 3 anions in roof run-off rainwater (Kingston-upon-Thames, Surrey, UK) stabilized by the addition of copper salt ($10 \, \mu g \, l^{-1}$)
CRM609	Groundwater – trace elements (low level)	Certified values for 5 metals in acidified (pH < 1.5) groundwater
CRM610	Groundwater – trace elements (high level)	Certified values for 5 metals in acidified (pH < 1.5) groundwater
LGC6019	River water (River Thames, UK) – metals	Certified values for 12 metals in acidified (pH 2) river water (Henley-on-Thames, Berkshire, UK)

Table 2.1 (*continued*)

Code	Name	Description
LGC6020	River water (River Thames, UK) – anions	Certified values for 3 anions in acidified (pH 2) river water (Henley-on-Thames, Berkshire, UK)
SPS-SW1	Surface water – trace metals	Certified values for 38 metals in surface water
SPS-SW2	Surface water – trace metals	Certified values for 44 metals in surface water
CRM505	Estuarine water – trace elements	Certified values for 4 metals in acidified (pH 1.5) estuarine water
SLEW-3	Estuarine water – trace elements	Certified values for 11 metals in estuarine water collected in San Francisco Bay, CA, USA, at a depth of 5 m
CRM403	Sea water – trace elements	Certified values for 6 metals in acidified (pH 1.5) sea water
NASS-5	Sea water – trace elements	Certified values for 10 metals in sea water collected in the North Atlantic at a depth of 10 m
LGC6175	Landfill leachate – trace elements	Certified values for 9 metals
LGC6176	Landfill leachate – anions	Certified values for 4 anions and chemical oxygen demand
SPS-WW1	Waste water – trace metals	Certified values for 13 metals
SPS-WW2	Waste water – trace metals	Certified values for 13 metals
SPS-NUTR-WW1	Waste water – anions	Certified values for 5 anions and total nitrogen
SPS-NUTR-WW2	Waste water – anions	Certified values for 5 anions and total nitrogen
Sediment		
CRM601	Sediment – extractable elements (three-step extraction)	Certified values for 5 metals
CRM320	River sediment – trace elements	Certified values for 10 metals
GBW08301	River sediment – trace elements	Certified values for 11 metals
SRM1939a	River sediment – PCBs and chlorinated pesticides	Certified values for 20 PCB congeners and 3 chlorinated pesticides collected from the Hudson River, New York State, USA

(*continued overleaf*)

Table 2.1   (*continued*)

Code	Name	Description
SRM1944	New York/New Jersey (USA) waterway sediment – PCBs and PAHs	Certified values for 35 PCBs, 24 PAHs, 4 chlorinated pesticides and 9 metals collected from six sites in the vicinity of New York Bay and Newark Bay, USA
WQB-1	Lake sediment – trace metals	Certified values for 5 metals
WQB-3	Lake sediment – trace metals	Certified values for 9 metals
DX-1	Lake sediment – dioxin and furan congeners	Certified values for dioxin and furan congeners
DX-2	Lake sediment – dioxin and furan congeners	Certified values for dioxin and furan congeners
EC-2	Lake sediment – organic contaminants	Certified values for 10 PAHs, 12 chlorobenzenes and hexachlorobutadiene collected from Hamilton Harbour and Lake Ontario, Canada
EC-3	Lake sediment – organic contaminants	Certified values for 6 PAHs, 6 chlorobenzenes and hexachlorobutadiene collected from Niagara River in Lake Ontario, Canada
CRM535	Freshwater harbour sediment – PAHs	Certified values for 7 PAHs
CRM536	Freshwater harbour sediment – PCBs	Certified values for 13 PCBs
Marine Sediment		
LGC6137	Estuarine sediment – extractable metals	Certified values for 19 metals collected from the Severn Estuary, UK
CRM462	Coastal sediment – organotin compounds	Certified values for tributyltin and dibutyltin
CRM580	Estuarine sediment – mercury and methylmercury	Certified values for mercury and methylmercury
IAEA-383	Marine sediment	Certified values for organochlorine compounds and PAHs
IAEA-408	Marine sediment	Certified values for organochlorine compounds and PAHs
SRM1646a	Estuarine sediment – metals	Certified values for 19 metals collected from Chesapeake Bay, MD, USA

Table 2.1 (*continued*)

Code	Name	Description
HS-1	Marine sediment – PCBs	Certified values for total and individual PCBs collected from Nova Scotia Harbour, Canada
HS-2	Marine sediment – PCBs	Certified values for total and individual PCBs collected from Nova Scotia Harbour, Canada
PACS-2	Harbour sediment – trace elements and organotin compounds	Certified values for 28 metals and mono-, di- and tributyltin compounds collected from Esquimalt Harbour, British Columbia, Canada
HS-3B; HS-4B; HS-5; HS-6	Harbour sediment – PAHs	Certified values for (16–21) PAHs collected from a harbour in Nova Scotia, Canada
MURST-ISS-A1	Antarctic sediment – trace elements	Certified values for 10 metals collected during the IX Italian Expedition (1993–1994) in Antarctica
Soil – uncontaminated		
CRM142R	Light sandy soil – trace elements	Certified values for 7 total and 4 aqua regia-soluble metals
CM17001	Light sandy soil – trace elements	Certified values for 11 total, 12 aqua regia-soluble, 10 boiling ($2\,mol\,l^{-1}$) nitric acid-soluble, and 11 cold ($2\,mol\,l^{-1}$) nitric acid-soluble metals
CM17002	Light sandy soil – trace elements	Certified values for 12 total, 11 aqua regia-soluble, 12 boiling ($2\,mol\,l^{-1}$) nitric acid-soluble, and 11 cold ($2\,mol\,l^{-1}$) nitric acid-soluble metals
CM17003	Silty clay loam soil – trace elements	Certified values for 11 total, 11 aqua regia-soluble, 11 boiling ($2\,mol\,l^{-1}$) nitric acid-soluble, and 11 cold ($2\,mol\,l^{-1}$) nitric acid-soluble metals
CM17004	Silty clay loam soil – trace elements	Certified values for 12 total, 11 aqua regia-soluble, 11 boiling ($2\,mol\,l^{-1}$) nitric acid-soluble, and 12 cold ($2\,mol\,l^{-1}$) nitric acid-soluble metals
CRM141R	Calcareous loam soil – trace elements	Certified values for 9 total and 9 aqua regia-soluble metals
CRM600	Calcareous loam soil – extractable trace elements	Certified values for 5 EDTA- and 2 DTPA-soluble metals

(continued overleaf)

Table 2.1 (*continued*)

Code	Name	Description
Soil – contaminated		
LGC6135	Soil (brick works) – leachable and total metals	Certified values for 14 total and 19 leachable metals collected from Hackney Brick Works, London
CRM481	Industrial soil – PCBs	Certified values for 8 PCBs
CRM524	Contaminated industrial soil – PAHs	Certified values for 8 PAHs and pentachlorophenol
CRM529	Industrial sandy soil – PCDDs and PCDFs	Certified values for PCDDs and PCDFs
CRM530	Industrial clay soil – PCDDs and PCDFs	Certified values for PCDDs and PCDFs
RTC401	Superfund soil – Toxicity Characteristic Leaching Procedure Organics	Certified values for *o*-cresol, total cresol, lindane, pentachlorophenol and trichlorophenol
RTC910	Soil – PCBs	Certified value for 'Arochlor 1242'
RTC911	Soil – PCBs	Certified value for 'Arochlor 1254'
RTC915	Soil – PCBs	Certified value for 'Arochlor 1260'
RTC916	Soil – PCBs	Certified value for 'Arochlor 1248'
RTC302; RTC303; RTC304; RTC305; RTC306	Soil – BTEX	Certified values for BTEX
RTC350; RTC351; RTC352; RTC353; RTC354; RTC356.	Soil – TPHs	Certified values for TPHs
RTC106; RTC114; RTC109; RTC110; RTC111; RTC113	Soil – semi-VOAs	Certified values for range of semi-VOAs (9–17)
RTC112	Soil – phenols	Certified values for 9 phenols
RTC105; RTC107	Soil – semi-VOAs and pesticides	Certified values for 26 semi-VOAs and pesticides
RTC803	Soil – herbicides	Certified values for 9 herbicides
RTC804; RTC805	Soil – pesticides	Certified values for 3–9 pesticides
RTC023; RTC024; RTC025; RTC026; RTC027; RTC028	Soil – metals	Certified values for 19–23 metals
CRM143R	Sewage sludge-amended soil – major and trace elements	Certified values for 8 total and 6 aqua regia-soluble metals
CRM483	Sewage sludge-amended soil – extractable trace elements	Certified values for 6 EDTA- and 6 acetic acid-soluble metals

Table 2.1 *(continued)*

Code	Name	Description
CRM484	Sewage sludge-amended (terra rossa) soil – extractable trace elements	Certified values for 5 EDTA- and 5 acetic acid-soluble metals
Sewage sludges		
LGC6181	Sewage sludge – leachable metals	Certified values for 12 metals collected from a city water treatment plant immediately after discharge from a digestion tank
LGC144R	Sewage sludge (domestic origin) – trace metals	Certified values for 9 total and 9 aqua regia-soluble metals
LGC145R	Sewage sludge (mixed origin) – trace metals	Certified values for 8 total and 5 aqua regia-soluble metals
LGC146R	Sewage sludge (industrial origin) – trace metals	Certified values for 9 total and 9 aqua regia-soluble metals
CRM088	Sewage sludge – PAHs	Certified values for 8 PAHs
RTC018	Raw sewage sludge – metals	Certified values for 23 metals
RTC029	Sewage sludge – metals	Certified values for 22 metals
SRM2781	Domestic sludge – metals	Certified values for 10 metals
SRM2782	Industrial sludge – leachable and total metals	Certified values for 10 metals collected from an industrial site in northern New Jersey, USA

[a]PCB, polychlorinated biphenyl; PAH, polycyclic (polynuclear) aromatic hydrocarbon; PCDD, polychlorinated dibenzo-*p*-dioxin; PCDF, polychlorinated dibenzofuran; EDTA, ethylenediaminetetraacetic acid; DTPA, diethylenetriaminepentaacetic acid; BTEX, benzene–toluene–ethylbenzene–xylene(s); TPH, total petroleum hydrocarbon; semi-VOAs, semi-volatile organic analytes.

DQ 2.2
Comment on the types, i.e. metals vs. organics, of certified reference materials presented in Table 2.1.

Answer

The range of CRMs available for environmental analyses is an expanding area of development for certifying bodies. You should note from Table 2.1 that while metals in environmental matrices are dominant there is an increasing number of CRMs for organic compounds, e.g. polynuclear aromatic hydrocarbons (PAHs). This trend is likely to continue for some time.

- *Analysis of reagent blanks.* Analyse reagents whenever the batch is changed or a new reagent is introduced. Introduce a minimum number of reagent blanks (typically 5% of the sample load); this allows reagent purity to be assessed and, if necessary, controlled and also acts to assess the overall procedural blank.

- *Calibration with standards.* A minimum number of standards should be used to generate the analytical curve, e.g. six or seven. Daily verification of the working curve should also be carried out by using one or more standards within the linear working range.

SAQ 2.2

Examine the graphs shown in Figure 2.4 and determine their linear working ranges.

- *Analysis of duplicates.* This allows the precision of the method to be reported.

- *Maintenance of control charts.* Various types of control charts can be maintained for standards, reagent blanks and replicate analytes. The purpose of each type of chart is to assess the longer-term performance of the laboratory, instrument, operator or procedure, based on a statistical approach.

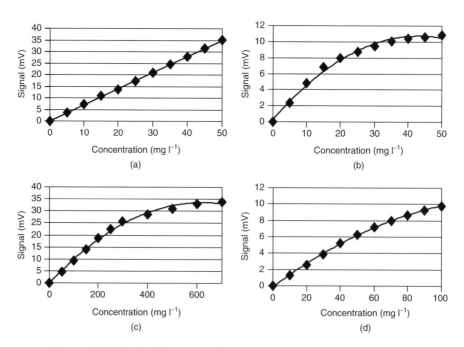

Figure 2.4 Plots for linear dynamic range determinations (cf. SAQ 2.2).

Summary

This chapter has considered the various aspects of planning the experimental work, introducing the concepts of quality assurance, accuracy and precision, and certified reference materials. It is now time to consider the initial phase of any environmental analysis procedure, i.e. sampling, and this will be discussed in the next chapter.

References

1. Anon, *Certified Reference Materials Catalogue*, Issue No. 3, Laboratory of the Government Chemist, Teddington, UK, 2000.

Chapter 3
Sampling

Learning Objectives

- To understand the concept of representative sampling.
- To understand the principles of sampling soil and sediment, water and air.
- To be able to determine the number of samples to be taken and be aware of the limitations.

3.1 Introduction

In an ideal world, all environmental samples would be analysed instantaneously without any need to transport samples to a laboratory. However, things are never quite like that. In order to economize on time, effort and cost, the areas under investigation must be sampled. Therefore, sampling of a particular site, lake or the atmosphere is required. Sampling constitutes the most important aspect of environmental analysis as without effective sampling all of the subsequent data generated are worthless. There are two primary types of sampling for environmental analysis, i.e. *random sampling* and *purposeful sampling*. The most important is random sampling as it infers no selectiveness to the sampling process.

DQ 3.1
Before reading on, how would you sample an agricultural field for soil samples, a flowing river for water samples, or the indoor air in an industrial manufacturing site?

Answer
Probably you have been able to suggest some approaches. Now read on to see if your ideas are appropriate.

Any sampling protocol involves the selection of the sample points, the size and shape of the sample area, and the number of sampling units in each sample. Before this can be done, information regarding the likely distribution of the contaminants under investigation is required. Contamination from inorganics or organics can be random, uniform (homogenous), patchy, stratified (homogenous within sub-areas) or present as a gradient (Figure 3.1). Preliminary testing of the site (a pilot study) is therefore beneficial to establish the likely distribution.

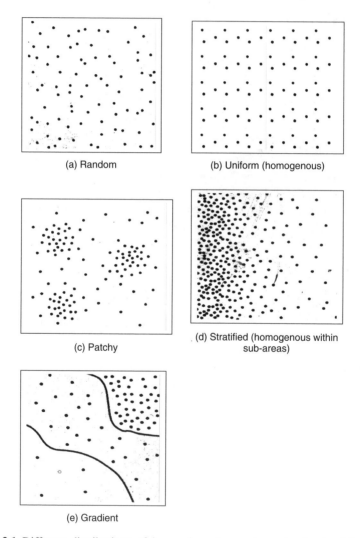

Figure 3.1 Different distributions of inorganic and organic contaminants: (a) random; (b) uniform (homogeneous); (c) patchy; (d) stratified (homogeneous within sub-areas); (e) gradient.

The key questions to be asked before sampling begins include the following:

1. Have arrangements been made to obtain samples from the site (e.g. permission from the site owner)?
2. Is specialized sampling equipment required and available?
3. How many samples and how many replicates are required?
4. Are the samples required for qualitative or quantitative analyses?
5. What chemical or physical tests are required?
6. What analytical methods and equipment are needed?
7. What mass/volume of sample is required for the analytical techniques to be used?
8. Is there a quality assurance protocol in place?
9. What types of container are required to store the samples and do you have enough available?
10. Do the containers require any pre-treatment/cleaning prior to use and has this been carried out?
11. Is any sample preservation required and do you know what it is?

3.2 Sampling Methods

The sampling position can be determined randomly, systematically or in a stratified random manner (Figure 3.2). In *random sampling*, a two-dimensional co-ordinate grid is superimposed on the area to be investigated. The selection of samples is completely 'down to the luck of the draw' without regard to the variation of the contaminant in the soil. It should be noted that the entire sample area is not sampled, but that every site on the grid has an equal chance of being selected for sampling. This type of sampling is ideal if the contaminant is homogeneous within the site.

Systematic sampling involves taking the position of the first sample at random and then taking further samples at fixed distances/directions from this. For example, samples may be taken at intervals of 5 m. This type of sampling has the potential to provide more accurate results than simple random sampling. However, if the soil contains a periodic (systematic) variation which coincides with this type of sampling, biased samples can result. An initial pilot study of the site can help prevent this.

Stratified sampling is commonly used in a location which is known to have contaminants heterogeneously distributed. This is therefore the most common approach to sampling. In this type of sampling, the site is sub-divided into smaller areas, each of which is fairly homogeneous, and thus more accurate sampling can take place. Each sub-area is then randomly sampled. The sub-dividing of the

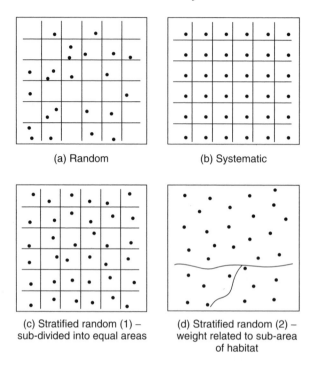

(a) Random	(b) Systematic

(c) Stratified random (1) – sub-divided into equal areas	(d) Stratified random (2) – weight related to sub-area of habitat

Figure 3.2 Basic methods used for sampling: (a) random; (b) systematic; (c) stratified random (1) – sub-division into equal areas; (d) stratified random (2) – weight related to sub-area of habitat.

site can be carried out either to give equal areas, or be related to known features within the site (see Figures 3.2(c) and 3.2(d), respectively).

3.3 Number of Samples

If the sampling site is homogeneous, then more samples need to be taken in order to achieve a certain accuracy. In reality, economic considerations often restrict both the quantity of material removed and the number of samples. In this case, the question becomes how many samples should be taken in order to achieve an acceptable error. This can be done, by using the following example, after first deciding on the magnitude of the error, E, that can be tolerated [1]:

$$E = \pm t (V)^{0.5} \tag{3.1}$$

$$V = S^2/n \tag{3.2}$$

where t is the test value, V the variance, S^2 the sum of the squares, and n the number of samples.

The sum of the squares, S^2, is calculated by using the following equation:

$$S = \sum (x_i - x)^2/(n - 1) \qquad (3.3)$$

The number of samples to be taken can then be calculated by using the following relationship:

$$n = t^2 S^2/E^2 \qquad (3.4)$$

SAQ 3.1

A series of 10 samples was taken at random from a site. The following results were obtained on the samples for lead. Using the above equations, calculate the magnitude of error that can be tolerated.

Sample	1	2	3	4	5	6	7	8	9	10
Pb (x_i) (ppm)	91	95	104	82	95	103	97	89	85	89
X (mean)	93	93	93	93	93	93	93	93	93	93
$(x_i - x)$	-2	2	11	-11	2	10	4	-4	-8	-4
$(x_i - x)^2$	4	4	121	121	4	100	16	16	64	16

SAQ 3.2

If the level of precision obtained from SAQ 3.1 is not acceptable and a maximum error of 2 ppm is only allowed, how many samples would need to be taken?

3.4 Sampling Soil and Sediment

Soil is an heterogenous material and significant variations (chemical and physical) can occur over even a small site, e.g. a field.

DQ 3.2

What might lead to different chemical and physical variations over the chosen site?

Answer

Variations can occur due to different topography, farming procedures, soil type, drainage and the underlying geology.

Obtaining a representative sample is therefore important (see Section 3.2 above). The tools required for sampling include an auger, a spade and a trowel. For shallow sampling of disturbed material, a trowel is sufficient to gather material. Once obtained, it should be placed in a polythene bag, sealed and clearly labelled with a permanent marker pen. Deep sampling can be carried out either by using augers or spades, by utilizing trenches and road banks, or by digging soil pits and exposing soil profiles.

An *auger* is a device that can be screwed into the ground to remove soil. The commonest types are the twin blade and the corkscrew (Figure 3.3). Both return disturbed soil to the surface. Augers are useful for pilot surveys of sites. The general procedure for using an auger is as follows:

1. Identify the site to be sampled.
2. Bang into the ground, using a mallet, a piece of heavy-walled PVC tubing. This will act as a guide for the auger.
3. Place the auger in the PVC tubing and turn the handle. This should allow the auger to collect soil. Once the auger is filled, remove from the ground and place the collected soil in a plastic bag. A sub-sample of the soil is then transferred into a pre-labelled and pre-cleaned glass jar. The lid is then placed on the jar and the sample stored on ice prior to returning to the laboratory.
4. Repeat Step 3, until the required depth has been achieved. (Note that care is needed with the 0–15 cm soil layer, as this can often be compacted.)

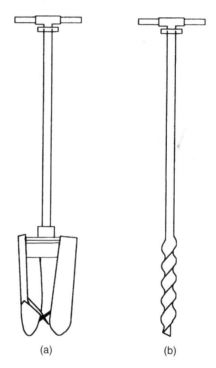

(a)   (b)

Figure 3.3 Types of augers used for soil sampling: (a) twin blade; (b) corkscrew.

Soil dug from a pit should be transferred to a plastic sheet placed on the ground – *not* directly onto the surrounding grassland. Always keep one side of the pit 'clean' by trimming with a trowel. This will allow the soil stratigraphy to be observed and recorded. In the vertical section, soils can be characterized by various layers of differing composition, known as a *soil profile*. A typical profile may consist of the following:

- The *L layer*: the litter layer, which is composed of the debris from plants and animals.
- The *A horizon* (top soil): the uppermost horizon, which contains mineral matter and some organic matter from the L layer.
- The *B horizon* (sub-soil): this lies below the A horizon and has a lower organic matter content.
- The *C horizon*: this is part of the underlying parent rock from which the soil above was derived.

Use a tape measure to allow the recording of depths and hence the thicknesses of the soil horizons in the profile. Colour is used to identify the different horizons within the profile. Standardized descriptions of colour can be obtained by the use of the Munsell Soil Colour Chart System (see Chapter 1).

When soil sampling has been completed, replace any unwanted soil and cover with grass sods (if appropriate). After sampling, it is common, after transportation of the sample back to the laboratory, to let the sample air-dry (at $<30°C$).

DQ 3.3

As you are trying to remove soil moisture, why not dry the soil in an oven?

Answer

The use of elevated temperatures can lead to loss of both volatile metals, e.g. mercury, and volatile organic species, e.g. BTEX (benzene–toluene–ethylbenzene–xylene(s)) compounds.

The air-dried sample is then sieved (2 mm diameter holes) to remove stones, large roots, etc., prior to sub-sampling to obtain a small representative sample of finely ground soil (the sample will often be sieved again to reduce the particle size). The purpose of this process is to obtain a sample suitable for the analytical technique and yet still remain representative of the original bulk sample. One of the most popular methods of sub-sampling the air-dried soil to obtain a representative sample for subsequent extraction/digestion and analysis is *coning* and *quartering*. In this procedure, the dried soil is thoroughly mixed and poured onto a clean sheet of polythene to form a cone. The latter is then divided into four quarters, using a cross made of sheet aluminium,

and the two opposite quarters combined and mixed to form a sub-sample of
approximately half the original weight. This half sample is again coned and quar-
tered and the process repeated until a sub-sample of the desired sample weight
is obtained.

3.5 Sampling Water

Water is a common substance to sample considering that the earth's surface is
composed of ca. 70% water.

DQ 3.4

How many different types or classifications of water can you think of?

Answer

*The different types of water can be classified as follows: surface waters
(rivers, lakes, run-off, etc.), groundwaters and spring waters, waste
waters (mine drainage, landfill leachate, industrial effluent, etc.), saline
waters, estuarine waters and brines, waters resulting from atmospheric
precipitation and condensation (rain, snow, fog, dew, etc.), process
waters, potable (drinking) waters, glacial melt waters, steam, water
for sub-surface injections, and water discharges, including water-borne
materials and water-formed deposits.*

While water would appear to be homogeneous, this is not in fact the case.
Water is heterogeneous, both spatially and temporally, thus making it extremely
difficult to obtain representative samples. Stratification within oceans, lakes and
rivers is common with variations in flow, chemical composition and temperature
all occurring. Variation with respect to time (temporal) can occur, for example,
because of heavy precipitation (snow, rain, etc.) and seasonal changes.

Water samplers can be either *automatic* or *manually operated*. Automatic samplers are used to collect samples at either fixed time-intervals or in proportion to
the flow and then to retain the water sample in a separate container. These are
commonly used, for example, in rivers or from a point source (effluent outfall). In
addition, automatic samplers can be used to allow the collection of time-averaged
samples or precipitation. In the case of the latter, the onset of rainfall triggers
the collection mechanism. Manually operated samplers are essentially open tubes
of known volume (typically 1 to 30 l) fitted with a closure mechanism at each
end. They are constructed of stainless-steel or PVC. Manually operated samplers
are particularly useful when sampling from open waters (e.g. oceans, seas, lakes,
etc.) at specific depths. The sampling device is lowered on a calibrated line to
the specific sampling depth, the sample is taken and then the top and bottom lids
are closed and sealed.

A good manual water sampler should possess the following attributes:

- provide a rapid descent in the water;
- be substantial enough to prevent minimal drift from the vertical;
- have a suitable closing and sealing mechanism to retain the sample but not allow the ingress of the surrounding water;
- be inert and hence not contaminate the sample;
- be easy to use and maintain;
- have an appropriate sample capacity.

A typical manual water-sampling device is shown in Figure 3.4.

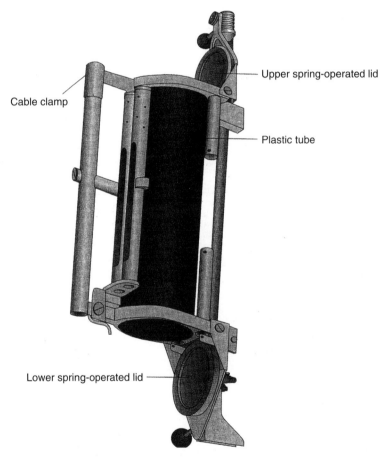

Figure 3.4 A typical manual device used for water sampling [2]. From Jones, A., Duck, R., Reed, R. and Weyers, J., *Practical Skills in Environmental Analysis*, Prentice Hall, Harlow, UK, 2000. © Pearson Education Limited 2000, reprinted by permission of Pearson Education Limited.

(a)

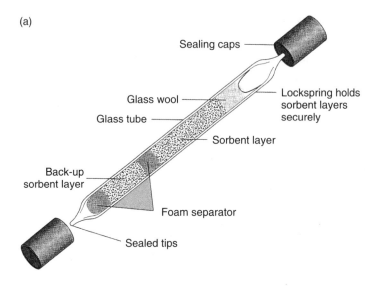

(b)

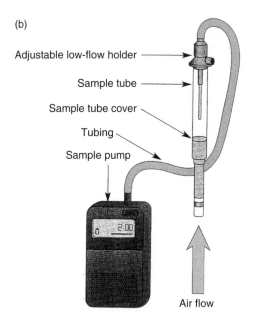

Figure 3.5 Air sampling: (a) a typical sorbent tube; (b) the system used to carry out measurements [2]. From Jones, A., Duck, R., Reed, R. and Weyers, J., *Practical Skills in Environmental Analysis*, Prentice Hall, Harlow, UK, 2000. © Pearson Education Limited 2000, reprinted by permission of Pearson Education Limited.

3.6 Sampling Air

The sampling of air can be classified into two groups, namely *particulate sampling*, in which particles are collected on filters, and *vapour/gas sampling*, in which air-borne compounds are trapped on a sorbent. Air sampling can take two forms. In the first, i.e. *passive sampling*, air-borne material diffuses onto filters and is retained. In the second type of sampling, the air-borne material is *actively* pumped through a filter or sorbent and then retained. A filter simply presents a physical barrier, while a sorbent provides an active site for chemical/physical retention of the material. The filters used can range in composition from fibre glass to cellulose fibres, while sorbents range from ion-exchange resins to polymeric substrates.

In sorbent tube sampling (Figure 3.5), volatile and semi-volatile compounds are pumped from the air and trapped on the surface of the sorbent. By sampling a measured quantity of air (typical volumes of $10-500 \text{ m}^3$), quantitative sampling is possible. The sorbent tube is then sealed and transported back to the laboratory for analysis. Desorption of volatile and semi-volatile compounds takes place either by the use of organic solvents (solvent extraction) or heat (thermal desorption), followed by analysis using gas chromatography (see later).

DQ 3.5

How would you sample snow and ice?

Answer

As the snow and ice are both frozen they should be treated in the same way as any other solid. You might therefore consider using an auger to sample them. You should also remember that unless you are doing this in 'cold climates' you might also need to sample the melt (liquid) fractions of both the snow and ice.

Summary

This chapter has focused on the different methods of sampling solids, liquids and gases. Upon its completion, you should feel confident enough to be able to sample any of these different forms and be ready to consider the storage conditions described in the next chapter.

References

1. Tan, K. H., *Soil Sampling, Preparation and Analysis*, Marcel Dekker, New York, 1996, p. 13.
2. Jones, A., Duck, R., Reed, R. and Weyers, J., *Practical Skills in Environmental Science*, Prentice Hall, Harlow, UK, 2000, pp. 102–103.

Chapter 4

Storage of Samples

Learning Objectives

- To understand the issues associated with the storage of samples.
- To understand the concept of sample storage.
- To appreciate the different methods available for sample preservation for metals and organics.
- To appreciate the difficulty associated with maintaining chemical species information, i.e. speciation.

4.1 Introduction

In an ideal situation, samples would be analysed *in situ* without the need for sampling, storage and transport to the laboratory, before the analysis. However, this is rarely the case. We have seen in the previous chapter that sampling is crucial to obtaining a representative sample. Once this has been achieved, it then becomes essential that the acquired sample is stored and transported to the laboratory in the same state as it was sampled.

DQ 4.1

What problems may occur when a sample is stored?

Answer

The concern with the storage of samples is that losses can occur, due to adsorption to the storage vessel walls, or that potential contaminants can enter the sample, from desorption or leaching from the storage vessels.

4.2 Methods

The problems can all lead to the analyst getting the wrong answer or at least an unexpected answer after the analysis has taken place. It is probably true to state that complete and unequivocal preservation of samples is impossible. However, the use of preservation methods should be able to retard any chemical or biological changes that will inevitably continue after sampling has occurred. The goal therefore is to store samples for the shortest possible time-interval between sampling and analysis. Indeed, in some instances where analytes are known to be unstable or volatile it may be necessary to perform the analysis immediately upon receipt, or even not at all! Methods of preservation are relatively few and are generally intended to fulfil the following criteria:

- to retard biological action
- to retard hydrolysis of chemical compounds and complexes
- to reduce volatility of constituents
- to reduce adsorption effects

Similarly, the preservation methods available are equally limited and constitute the following approaches:

- pH control
- addition of chemicals
- refrigeration
- freezing

It is important to note how long the sample has been stored and under what conditions storage has been carried out.

DQ 4.2

Why might the nature and type of storage vessel be important?

Answer

For example, if it is known that the analyte is light sensitive it is then essential that the sample is stored in a brown glass container to prevent photochemical degradation. For volatile species, it is also desirable that the sample is stored in a well-sealed container. In most cases, the use of glass containers is recommended as there is little opportunity for contamination to result as a consequence of the vessel itself. It is also important that the appropriate sized container is used. It is better to completely fill the storage container rather than leave a significant headspace above the sample. This acts to reduce any oxidation that may occur. In addition to glass containers, polyethylene or polytetrafluoroethylene (PTFE) containers are appropriate to use for solid samples. Plastic containers are not recommended for aqueous samples as plasticizers, e.g. phthalates, are prone to leach from the vessels which can cause problems at later stages of the analysis.

Whatever method of storage is chosen it is desirable to perform experiments to identify that the analyte of interest does not undergo any chemical or microbial degradation and that contamination is kept to a minimum. Some selected examples of methods of preservation for water samples are shown in Table 4.1.

SAQ 4.1

How would you preserve aqueous samples for the analysis of (a) total lead, (b) sulfate, and (c) dieldrin?

A recent review has highlighted the stability of chemical species (in speciation studies) with respect to environmental matrices [1]. These results are summarized in Table 4.2. It should be noted that samples must be analysed as quickly as possible after collection. The times given in this table are the *maximum* times that samples should be stored before analysis. It should also be noted that the guidelines given are general in nature. For example, many organophosphorus pesticides (not shown in Table 4.2) can be preserved by the addition of hydrochloric acid. However, as an exception to this, the (organophosphorus) pesticide diazinon breaks down when acidified.

It is also important to *prepare* the sample container prior to sample storage. In the case of samples for metal analysis, the following procedure is recommended. The sample container, borosilicate glass or plastic (polyethylene, polypropylene or 'Teflon' (PTFE)) should be treated in the following sequence:

1. Wash in detergent to remove any solid residues (may not be necessary).

2. Soak containers for at least 24 h in an acid bath (10% nitric acid; 10 ml of concentrated nitric acid in every 100 ml volume of water).

3. Rinse with deionized distilled water.

4. Repeat the rinse step at least twice more with deionized distilled water.

In some cases, the removal of organic residues from glassware requires the use of a chromic acid wash (prepared by adding 100 ml of concentrated sulfuric acid slowly and with constant stirring to a solution of 5 g of sodium dichromate in 5 ml of water). In this situation, it is important to thoroughly rinse the glassware in deionized water so as to remove any trace of chromium, particularly if the latter species is part of the analytical scheme.

Table 4.1 Selected examples of preservation techniques for water samples[a]

Compound	Container	Preservation	Maximum holding time
Metals			
Total	Polyethylene, with a polypropylene cap (no liner), or glass	100 ml of water acidified (HNO_3) to pH < 2	6 months
Dissolved	Polyethylene, with a polypropylene cap (no liner), or glass	200 ml of water filtered on site, then acidify (HNO_3) to pH < 2	6 months
Suspended	Polyethylene, with a polypropylene cap (no liner), or glass	200 ml of water filtered on site	6 months
Chromium	Polyethylene, with a polypropylene cap (no liner), or glass	Cool to 4°C (200 ml of water)	24 h
Non-metals			
Fluoride	Plastic or glass	None required (300 ml of water)	28 days
Chloride	Plastic or glass	None required (200 ml of water)	28 days
Bromide	Plastic or glass	None required (100 ml of water)	28 days
Nitrate and nitrite	Plastic or glass	Cool to 4°C, add H_2SO_4 to pH < 2 (100 ml of water)	28 days
Sulfate	Plastic or glass	Cool to 4°C (50 ml of water)	28 days
Organics			
Pesticides (organochlorine)	Glass	1 ml of a 10 mg ml^{-1} $HgCl_2$ or adding of extraction solvent (500 ml of water)	7 days, 40 days after extraction
Pesticides (organophosphorus)	Glass	1 ml of a 10 mg ml^{-1} $HgCl_2$ or adding of extraction solvent (500 ml of water)	14 days, 28 days after extraction
Pesticides (chlorinated herbicides)	Glass	Cool to 4°C, seal, add HCl to pH < 2 (500 ml of water)	14 days
Pesticides (polar)	Glass	1 ml of a 10 mg ml^{-1} $HgCl_2$ (500 ml of water)	28 days
Phenolic compounds	Glass	Cool to 4°C, add H_2SO_4 to pH < 2 (500 ml of water)	28 days
Biological oxygen demand (BOD)	Plastic or glass	Cool to 4°C (1000 ml of water)	48 h
Chemical oxygen demand (COD)	Plastic or glass	Cool to 4°C, add H_2SO_4 to pH < 2 (50 ml of water)	28 days

[a] As recommended by different agencies, e.g. The Environmental Protection Agency (EPA) and The International Organization for Standardization (ISO).

Table 4.2 Stability of chemical species in environmental matrices [1]

Matrix	Container	Preservation	Maximum holding time	Reference
Organotin species[a]				
Synthetic solutions (TPT and TBT)	Polyethylene	HCl acidified water, in the dark, at 4°C	3 months	2
Synthetic solutions (TPT and TBT)	Brown glass	HCl acidified water at 25°C	20 days	3
Sea water (TBT)	Pre-washed 'Pyrex' bottles	Filtered sample acidified to pH < 2 and stored at 4°C	—	4
Sea water (butyltins but not TPT)	Polycarbonate bottles	Cooled to 4°C and stored in the dark	7 months	5
Sea water	C$_{18}$ cartridges	Room temperature	60 days for TPT and 7 months for butyltins	5
Sediment (wet) (TBT)	—	Cooled to 4°C or frozen, followed by different drying procedures, i.e. oven drying at 50°C, freeze-drying and air-drying. NOT suitable for DBT and MBT	4 months	4
Sediment (wet) (OT)	—	Loss of butyltins when using either air-drying under the action of light, an IR lamp or oven drying at 110°C. Butyltin and phenyltin species unaffected by lyophilization or desiccation procedures	3 months for phenyltins and 1 year for butyltins	6
Freshwater sediment (OT)	—	Store at −20°C, independent of the treatment used for preservation	18 months	5
Oysters and cockles	—	Lyophilization (drying procedure) allows butyltin species to be stable if stored at −20°C and in the dark	150 days	5
Mussels	—	Lyophilization (drying procedure) allows butyltin species to be stable if stored at −20°C and in the dark	44 months	7
Organoarsenicals[b]				
Distilled and natural (As(III) and As(V))	'Pyrex' and polyethylene bottles	Sulfuric (0.2 vol%) acidified water (pH < 1.5) at room temperature; 40% losses if pH increased	125 days	8

(continued overleaf)

Table 4.2 (*continued*)

Matrix	Container	Preservation	Maximum holding time	Reference
Interstitial water (As(III))	'Pyrex' and polyethylene bottles	HCl acidified water (pH 2) and cool to ~0°C	6 weeks	9
Water (methylated species)	—	HCl (4 ml l⁻¹ of sample) acidified water	Several months	10
River water (As(V), MMAA and DMAA)	—	No preservation	20–26 h	11
Marine samples	—	Freeze at −20°C followed by freeze-drying	—	12–14
Mercury species[c]				
Water	Glass bottles	—	—	15, 16
Water (total Hg)	Poly(ethylene terephthalate)	0.5% of 20% (wt/vol) potassium dichromate dissolved in 1:1 HNO_3	10 days	17
Water	—	Presence of humic substances will preserve methylmercury if stored at 4°C and in the dark	33 days	18
Fish extract solutions (fish homogenate mixed with $HCl/CuSO_4$ and extracted into toluene)	—	Store at 4°C in the dark	5 months	19
Tuna fish (reference material)	—	Store at 20 and 40°C in the dark	12 months	20
Chromium species[d]				
Waste water	—	—	As soon as possible	21
Water (Cr(III) and Cr(VI))	Quartz ampoules	Store at 5°C at pH 6.4 (with HCO_3^-/H_2CO_3 buffer under CO_2 to avoid Cr(VI) reduction)	228 days	22

[a] For example, organotins (OTs), monobutyltin (MBT), dibutyltin (DBT), tributyltin (TBT) and triphenyltin (TPT).
[b] For example, arsenite (As(III)), arsenate (As(V)), monomethylarsonic acid (MMAA), dimethylarsinic acid (DMAA), arsenobetaine (AsB) and arsenocholine (AsC).
[c] For example, methylmercury.
[d] For example, Cr(III) and Cr(VI).

4.3 Summary

It has been seen that the main emphasis in this chapter is to store the sample for the minimum amount of time. If the sample has to be stored because the next stage of the procedure cannot be carried out straight away due to lack of equipment, staff availability or because the sample must be transported over a large distance, procedures are available and have been discussed. The following chapters next describe how the sample is to be treated prior to the analysis step. This frequently involves extraction from the matrix for organic compounds, while for metals digestion of the matrix is often used although, as you will observe, other selective approaches are possible.

References

1. Gomez-Ariza, J. L., Morales, E., Sanchez-Rodas, D. and Giraldez, I., *Trends Anal. Chem.*, **19**, 200–209 (2000).
2. Quevauviller, Ph., Astruc, M., Ebdon, L., Muntau, H., Cofino, W., Morabito, R. and Griepink, B., *Mikrochim. Acta.*, **123**, 163–173 (1996).
3. Bergmann, K., Rohr, U. and Neldhart, B., *Fresenius' J. Anal. Chem.*, **349**, 815–819 (1994).
4. Quevauviller, Ph. and Donard, O. F. X., *Fresenius' J. Anal. Chem.*, **339**, 6–14 (1991).
5. Gomez-Ariza, J. L., Giraldez, I., Morales, E., Ariese, F., Cofino, W. and Quevauviller, Ph., *J. Environ. Monit.*, **1**, 197–202 (1999).
6. Gomez-Ariza, J. L., Morales, E., Beltran, R., Giraldez, I. and Ruiz-Benitez, M., *Quim. Anal.*, **13**, S76–S79 (1994).
7. Morabito, R., Muntau, H., Corifino, W. and Quevauviller, Ph., *J. Environ. Monit.*, **1**, 75–82 (1999).
8. Cheam, V. and Agemian, H., *Analyst*, **105**, 737–743 (1980).
9. Aggett, J. and Kriegman, M. R., *Analyst*, **112**, 153–157 (1987).
10. Crescelius, E. A., Bloom, N. S., Cowan, C. E. and Jenne, E. A., *Speciation of Selenium and Arsenic in Natural Waters and Sediments*, Vol. 2, *Arsenic Speciation*, Electric Power Research Institute (EPRI), Batelle Northwest Laboratories, Washington, DC, 1986.
11. Anderson, R. K., Thomson, M. and Culbard, E., *Analyst*, **111**, 1153–1158.
12. Gailer, K. A., Francesconi, K. A., Edmonds, J. S. and Irgolic, K. J., *Appl. Organomet. Chem.*, **9**, 341–355 (1995).
13. Jitoh, F., Imura, H. and Suzuki, N., *Anal. Chim. Acta*, **228**, 85–91 (1990).
14. Goessler, W., Maher, W., Irgolic, K. J., Kuehnelt, D., SclagenHaufen, C. and Kaise, T., *Fresenius' J. Anal. Chem.*, **359**, 434–437 (1997).
15. American Public Health Association–American Water Works Association–Water Environment Association (APHA–AWWA–WEF), *Standards Methods for the Examination of Water and Wastewater*, 20th Edn, APHA–AWWA–WEF, Washington, DC, 1998.
16. Department of the Environment, Standing Committee of Analysts, *Methods for the Examination of Waters and Associated Materials*, Her Majesty's Stationery Office, London, 1985.
17. Copeland, D. D., Facer, M., Newton, R. and Walker, P. J., *Analyst*, **121**, 173–176 (1996).
18. Reinholdsson, F., Briche, C., Emteborg, H., Baxter, D. C. and Frech, W., 'Determination of mercury species in natural waters', in *Proceedings of Colloquium Analytische Atomsprektoskopie, 1995, (CANAS'95)*, Konstanz, Germany, April 2–7, 1995, Vol. 8, Welz, B. (Ed.), Bodenseewerk Perkin Elmer GmbH, Ueberlingen, Germany, 1996, pp. 339–405.
19. Quevauviller, Ph., Drabaek, I., Muntau, H. and Griepink, B., *Appl. Organomet. Chem.*, **7**, 413–420 (1993).
20. Quevauviller, Ph., Drabaek, I., Muntau, H., Bianchi, M., Bortoli, A. and Griepink, B., *Trends Anal. Chem.*, **15**, 390–397 (1996).
21. Pantsar-Kallio, M. and Manninen, P. K. G., *J. Chromatogr., A*, **750**, 89–95 (1996).
22. Dyg, S., Cornelius, R., Griepink, B. and Quevauviller, Ph., *Anal. Chim. Acta*, **286**, 297–308 (1994).

Sample Preparation for Inorganic Analysis

Chapter 5
Solids

Learning Objectives

- To appreciate the different approaches available for the preparation of solid samples for elemental analysis.
- To be able to carry out acid digestion (hot-plate and microwave) in a safe and controlled manner.
- To be aware of other decomposition methods (e.g. fusion and dry ashing).
- To understand the importance of chemical speciation studies.
- To appreciate the importance of chemical species identification with respect to mercury, tin, arsenic and chromium.
- To be able to carry out methylmercury extraction in a safe and controlled manner.
- To be able to carry out organotin extraction in a safe and controlled manner.
- To be able to carry out organoarsenical extraction in a safe and controlled manner.
- To be able to carry out hexavalent chromium extraction and analysis in a safe and controlled manner.
- To understand the relevance of selective extraction methods for soil studies.
- To be able to carry out single extraction procedures using EDTA, acetic acid and DTPA in a safe and controlled manner.
- To be able to carry out a sequential extraction procedure on a soil sample in a safe and controlled manner.
- To be able to carry out a simulated gastro-intestinal extraction of foodstuffs in a safe and controlled manner.
- To be able to carry out a physiologically based extraction test on soil in a safe and controlled manner.

5.1 Introduction

Analysis for metals in solids can be carried out by two different approaches, namely direct analysis of the solid, or after decomposition of the matrix to liberate the metal. Samples can be analysed directly for metals by using, for example, X-ray fluorescence (XRF) spectroscopy (see Chapter 11). This present chapter principally focuses on methods of decomposition of the matrix to liberate its metal content. In addition, selective methods of metal extraction are considered, together with appropriate methods of analysis.

5.2 Decomposition Techniques

Decomposition involves the liberation of the analyte (metal) of interest from an interfering matrix by using a reagent (mineral/oxidizing acids or fusion flux) and/or heat. The utilization of reagents (acids) and external heat sources can in itself cause problems. In elemental analysis, these problems are particularly focused on the risk of contamination and loss of analytes. It should be borne in mind that complete digestion may not always be required as atomic spectroscopy frequently uses a hot source, e.g. flame or inductively coupled plasma, which provides a secondary method of sample destruction. Therefore, methods that allow sample dissolution may equally be as useful.

DQ 5.1

Consider the difference between the terms 'digestion' and 'dissolution'.

Answer

Digestion *infers the complete destruction of the sample matrix whereas* **dissolution** *considers the liberation from the matrix of the analyte of interest (without the requirement for complete destruction of the matrix). In the latter case, it may still be possible to identify 'organic' parts of the matrix by using appropriate techniques.*

5.3 Dry Ashing

Probably the simplest of all decomposition systems involves the heating of the sample in a silica or porcelain crucible in a muffle furnace in the presence of air at 400–800°C. After decomposition, the residue is dissolved in acid and transferred to a volumetric flask prior to analysis. This allows organic matter to be destroyed. However, the method may also lead to the loss of volatile elements, e.g. Hg, Pb, Cd, Ca, As, Sb, Cr and Cu. Thus, while compounds can be added to retard the loss of volatiles, its use is limited. Due to the disadvantages of this method, namely:

- losses due to volatilization
- resistance to ashing by some materials
- difficult dissolution of ashed materials
- high risk of contamination

it has largely been replaced by *wet ashing*.

5.4 Acid Digestion (including the Use of Microwaves)

Acid digestion involves the use of mineral or oxidizing acids and an external heat source to decompose the sample matrix. The choice of an individual acid or combination of acids is dependent upon the nature of the matrix to be decomposed. The most obvious example of this relates to the digestion of a matrix containing silica (SiO_2), e.g. in a geological sample. In this situation, the only appropriate acid to digest the silica is hydrofluoric acid (HF). No other acid or combination of acids will liberate the metal of interest from the silica matrix.

SAQ 5.1

Why should HF be so effective for the digestion of silica?

A summary of the most common acids types used and their applications is shown in Table 5.1.

Table 5.1 Some examples of common acids used for wet decomposition[a]

Acid	Boiling point (°C)	Comments
Hydrochloric (HCl)	110	Useful for salts of carbonates, phosphates, some oxides and some sulfides. A weak reducing agent; not generally used to dissolve organic matter
Sulfuric (H_2SO_4)	338	Useful for releasing a volatile product; good oxidizing properties for ores, metals, alloys, oxides and hydroxides; often used in combination with HNO_3. **Caution**: H_2SO_4 must never be used in PTFE vessels (PTFE has a melting point of 327°C and deforms at 260°C)
Nitric (HNO_3)	122	Oxidizing attack on many samples not dissolved by HCl; liberates trace elements as the soluble nitrate salt. Useful for the dissolution of metals, alloys and biological samples

(*continued overleaf*)

Table 5.1 (*continued*)

Acid	Boiling point (°C)	Comments
Perchloric ($HClO_4$)	203	At fuming temperatures, a strong oxidizing agent for organic matter. **Caution**: violent, explosive reactions may occur – care is needed. Samples are normally pre-treated with HNO_3 prior to addition of $HClO_4$
Hydrofluoric (HF)	112	For digestion of silica-based materials; forms SiF_6^{2-} in acid solution. **Caution** is required in its use; glass containers should not be used, only plastic vessels. In case of spillages, calcium gluconate gel (for treatment of skin contact sites) should be available prior to usage; evacuate to hospital immediately if skin is exposed to liquid HF
Aqua regia (nitric/hydrochloric)	—	A 1:3 vol/vol mixture of HNO_3:HCl is called aqua regia; forms a reactive intermediate, NOCl. Used for metals, alloys, sulfides and other ores – best known because of its ability to dissolve Au, Pd and Pt

[a]Protective clothing/eyewear is essential in the use of concentrated acids. All acids should be handled with care and in a fume cupboard.

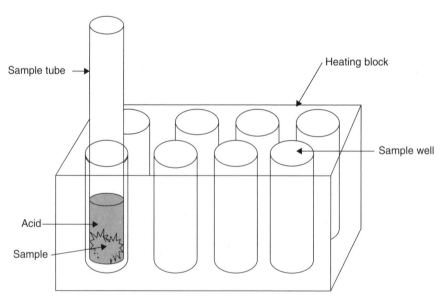

Figure 5.1 Schematic of a commercial acid digestion system.

Once the choice of acid is made, the sample is placed into an appropriate vessel for the decomposition stage. The choice of vessel, however, depends upon the nature of the heat source to be applied. Most commonly, the acid digestion of solid matrices has been carried out in open glass vessels (beakers or boiling tubes) using a hot-plate or multiple-sample digestor. The latter allows a number of boiling tubes (6, 12 or 24 tubes) to be placed into the well of a commercial digestor (Figure 5.1). In this manner, multiple samples can be simultaneously digested. The US Environmental Protection Agency (EPA) methods for the acid digestion of sediments, sludges and soils are outlined in Figures 5.2 and 5.3.

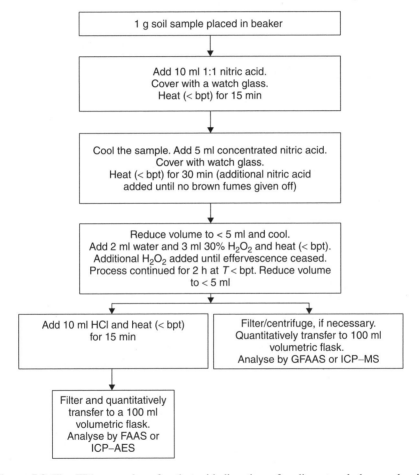

Figure 5.2 The EPA procedure for the acid digestion of sediments, sludges and soils using a hot-plate: GFAAS, graphite-furnace atomic absorption spectroscopy; FAAS, flame atomic absorption spectroscopy; ICP–MS, inductively coupled plasma–mass spectrometry; ICP–AES, inductively coupled plasma–atomic absorption spectroscopy [1].

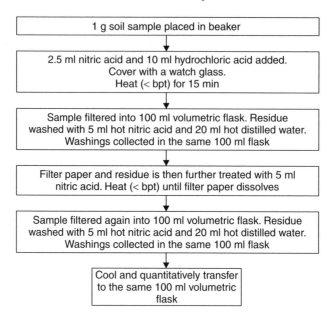

Figure 5.3 The EPA procedure for the acid digestion of sediments, sludges and soils, involving a hot-plate, used for the determination of Sb, Ba, Pb and As [2].

The digestion of foodstuffs can be carried out in a similar manner to that described for sediments, sludges and soils. A typical procedure for the digestion of foodstuffs, e.g. cereals, meats, fish and vegetables (excluding root vegetables) for the determination of total metal content (e.g. cadmium, copper, iron, lead and zinc) is described in Figure 5.4. If the level of metal in the sample is below the detection limit of the analytical technique being used it is necessary to pre-concentrate the metal present in the digest. One approach to this is the application of chelation–extraction using ammonium pyrrolidine dithiocarbamate–diethylammonium diethyldithiocarbamate (APDC–DDDC) into 4-methylpentan-2-one (MIBK) (see also Chapter 6). In this procedure, a $10 \text{ g} \text{l}^{-1}$ concentration of APDC with DDDC is used as a 1 vol% aqueous solution. The procedure adopted is as follows: 1 ml of the APDC–DDDC solution is added to 80 ml of sample digest in a suitable sample container and mixed on a vortex mixer for 10 s. Then, the metal chelate is extracted by the addition of 10 ml of MIBK and further vortex mixing for 20 s. After allowing this resultant mixture to stand for 5 min, the organic layer (MIBK containing the metal chelate) is removed and analysed. Metal standards for quantitation should be prepared in the same manner, i.e. chelated and extracted.

An alternative approach to conventional heating involves the use of microwave heating.

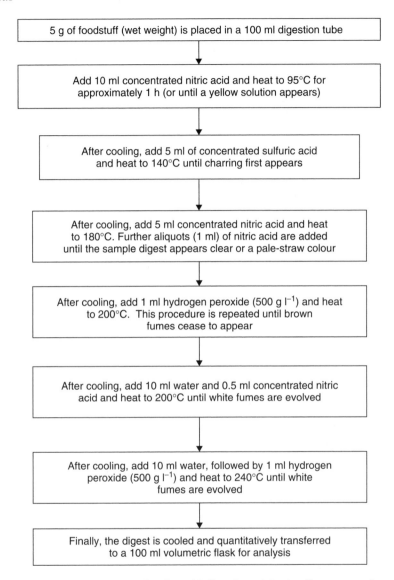

Figure 5.4 A typical procedure for the acid digestion of foodstuffs, e.g. cereals, meats, fish and vegetables, used for the determination of total metal content [3].

5.4.1 Microwave Digestion

The first reported use of a microwave oven for the acid digestion of samples for metal analysis was in 1975 [4]. Advances in technology by a variety of manufacturers mean that today there are two types of microwave heating systems commercially available, i.e. an open-focused and a closed-vessel system (for background information on microwave heating, see Box 5.1). In the open-style system, up to six sample vessels are heated simultaneously. A typical commercial system is the 'Simultaneous Temperature Accelerated Reaction' (STAR™) system from the CEM Corporation, USA. A schematic diagram of an open-focused

Box 5.1 Microwave Heating

Historical Background

Dr Percy Spencer, a scientist with Raytheon Corporation, USA, was working on a radar-related project in 1946 when he noted something unusual. He was testing a new vacuum tube, called a *magnetron*, when he discovered that the candy bar in his pocket had melted. He went on to try some popcorn, which when placed close to the magnetron cracked and popped. From this curious beginning was discovered the microwave oven of today. In 1947, the first commercial microwave ovens ('radarange') for heating food appeared in the marketplace. These ovens were both very large and expensive. However, developments over the years have meant that both the price and size have been reduced considerably.

Microwave Interaction with Matter

Microwaves are high-frequency electromagnetic radiation with a typical wavelength of 1 mm to 1 m. Many microwaves systems, both industrial and domestic, operate at a wavelength of around 12.2 cm (or a frequency of 2.45 GHz) to prevent interference with radio transmissions [5]. Microwaves are split into two parts, i.e. the electric-field component and the magnetic-field component. These are perpendicular to each other and the direction of propagation (travel) and vary sinusoidally. Microwaves are comparable to light in their characteristics. They are said to have particulate character as well as acting like waves. The 'particles' of microwave energy are known as *photons*. These photons are absorbed by the molecule in the lower-energy state (E_0) and the energy raises an electron to a higher-energy level (E_1). Since electrons occupy definite energy levels, changes in these levels are discrete and therefore do not occur continuously. The energy is said to be *quantized*. Only charged particles are affected by the electric-field part of the microwave. The Debye equation for the dielectric constant of a material determines the polarizability of the molecule. If these charged particles or polar molecules are free to move, this causes a current in the material. However, if they are bound strongly within the compound and consequently

Continued on page 57

Continued from page 56

are not mobile within the material, a different effect occurs. The particles re-orientate themselves so they are in-phase with the electric-field. This is known as *dielectric polarization* [6].

The latter is split into four components, with each being based upon the four different types of charged particles that are found in matter. These are electrons, nuclei, permanent dipoles and charges at interfaces. The total dielectric polarization of a material is the sum of all four components:

$$\alpha_1 = \alpha_e + \alpha_a + \alpha_d + \alpha_i$$

where α_1 is the total dielectric polarization, α_e is the electronic polarization (polarization of electrons round the nuclei), α_a is the atomic polarization (polarization of the nuclei), α_d is the dipolar polarization (polarization of permanent dipoles in the material), and α_i is the interfacial polarization (polarization of charges at the material interfaces).

The electric field of the microwaves is in a state of flux, i.e. it is continually polarizing and depolarizing. These frequent changes in the electric field of the microwaves cause similar changes in the dielectric polarization. Electronic and atomic polarization and depolarization occur more rapidly than the variation in the electric field, and have no effect on the heating of the material. Interfacial polarization (also known as the Maxwell–Wagner effect) only has a significant effect on dielectric heating when charged particles are suspended in a non-conducting medium, and are subjected to microwave radiation. The time-period of oscillation of the permanent dipoles is similar to that of the electric field of microwaves. The resulting polarization lags behind the reversal of the electric field and causes heating in the substance. These phenomena are thought to be the main contributors to dielectric heating.

Heating Methods

A possible reason for the reduced extraction times when using microwaves can be attributed to the different heating methods employed by microwave and conventional heating. The different heating profiles (Figure 5.5) obtained for water in a microwave and when using conventional methods show that liquid heated in a microwave reaches its boiling point much more rapidly than under conventional methods. In conventional heating, e.g. with a hot-plate, a finite period of time is required to heat the vessel before the heat is transferred to the solution. Thermal gradients are set up in the liquid due to convection currents. This means that only a small fraction of the liquid is at the required temperature.

Continued on page 58

Continued from page 57

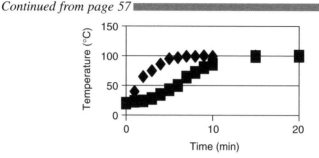

Figure 5.5 Heating profiles for deionized water: ■, conventional heating; ◆, microwave heating (cf. Box 5.1).

Microwaves heat the solution directly, without heating the vessel, and hence temperature gradients are kept to a minimum. Therefore, the rate of heating when using microwave radiation is faster than with conventional methods. Energy is not lost due to unnecessary heating of the vessel. Localized *superheating* can also occur [7].

Choice of Reagents

A substance that absorbs microwave energy strongly is called a *sensitizer*. The latter preferentially absorbs the radiation and passes the energy on to other molecules. Polar molecules and ionic solutions (usually acids) will absorb microwave energy strongly in relation to non-polar molecules. This is because they have a permanent dipole moment that will be affected by the microwaves. If extraction between non-polar molecules is required, then the choice of solvent is the main factor to consider. If the solvent molecule is not sensitive enough to the radiation, then there will be extraction. This is because the substance will not heat up.

Solvent Effects

A correct choice of solvent for microwave-assisted extraction is essential. The solvent must be able to absorb microwave radiation and pass it on in the form of heat to other molecules in the system. The following equation [8] measures how well a certain solvent will pass on energy to others:

$$\varepsilon''/\varepsilon' = \tan \delta$$

where δ is the dissipation factor, ε'' is the dielectric loss (a measure of the efficiency of conversion of microwave energy into heat energy), and ε' is

Continued on page 59

Continued from page 58

the dielectric constant (a measure of the polarizability of a molecule in an electric field).

Polar solvents, such as water, acetone and methanol, all readily absorb microwaves and are heated up when subjected to microwave radiation. Non-polar solvents, such as hexane and toluene, do not heat up when they are subjected to microwave irradiation.

microwave system is shown in Figure 5.6. The sample and acid (sulfuric acid can be used) are introduced into a glass container, which has the appearance of a large boiling/test tube, and is then fitted with a condensor to prevent loss of volatiles. The sample container is placed within the microwave cavity and heated.

A common commercial closed system is the Microwave Accelerated Reaction System (MARS™) 5, as supplied by the CEM Corporation, USA (Figure 5.7). This system allows up to 14 extraction vessels (XP-1500 Plus™) to be irradiated simultaneously. In addition, other features include a function for monitoring both pressure and temperature, and most notably, the system is equipped with

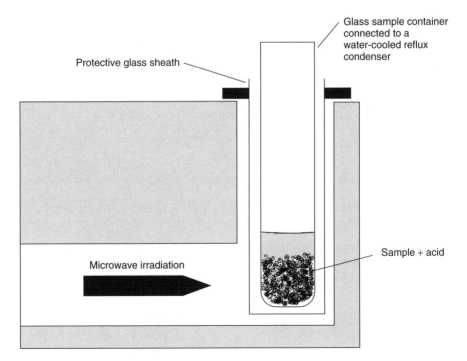

Figure 5.6 Schematic of an atmospheric, open-focused microwave digestion system.

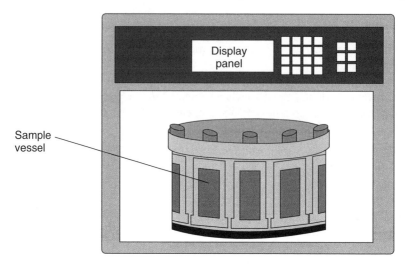

Figure 5.7 Schematic of a pressurized microwave digestion system.

an alarm to call attention to any unexpected release of flammable and toxic material. The microwave energy output of this system is 1500 W at a frequency of 2450 MHz at 100% power. Pressure (up to 800 psi) is continuously measured (measurements taken at the rate of 200 s^{-1}), while the temperature (up to 300°C) is monitored for all vessels every 7 s. All of the sample vessels are held in a carousel which is located within the microwave cavity. Each vessel has a vessel body and an inner liner. The liner is made of 'TFM' fluoropolymer and has a volume of 100 ml. A patented safety system (AutoVent Plus^{TM}) allows the venting of excess pressure within each extraction vessel. The system works by lifting of the vessel cap to release excess pressure and then immediately resealing to prevent loss of sample. If solvent leaking from the extraction vessel(s) does occur, a solvent monitoring system will automatically shut off the magnetron, while still allowing the exhaust fan to continue working.

DQ 5.2

What advantage in terms of digestion time could the use of a microwave system offer over a conventional hot-plate?

Answer

It would be faster using a microwave oven (consider the domestic microwave oven and a conventional electric or gas cooker), particularly if the digestion is carried out in sealed vessels (pressure and temperature effects).

The use of open vessels for digestion can lead to additional problems associated with loss by volatilization of element species. This can be rectified by the correct choice of reagents and the type of digestion apparatus to be used.

5.4.2 Microwave Digestion Procedure

Unless the microwave system used is capable of temperature feedback control, thus allowing it to replicate effectively the desired temperature profile, then the system will need to be calibrated. The procedure for calibration of microwave equipment is shown in Box 5.2. The typical operating procedure for digestion of a sample of sediment, sludge or soil is shown in Figure 5.8, while for siliceous and organically based matrices, see Figure 5.9. However, it should be borne in mind that all digestion vessels and volumetric flasks to be used should be acid cleaned prior to use in order to reduce the risk of contamination. In addition to

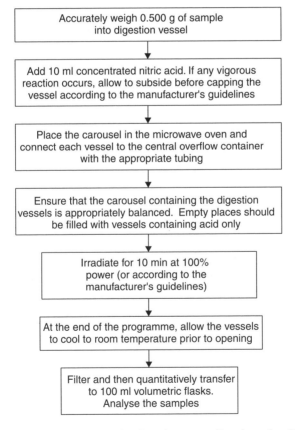

Figure 5.8 A typical procedure used for the microwave digestion of sediments, sludges and soils [9].

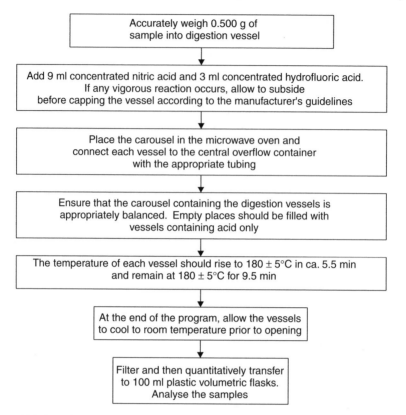

Figure 5.9 A typical procedure used for the microwave digestion of siliceous and organically based matrices [10].

the typical acid soak in a 10% nitric acid bath followed by rinsing in distilled water for polymeric or glass volumetric flasks, the following is suggested for the digestion vessels–leach with hot, i.e. >80°C, but <bpt (1:1 vol/vol) hydrochloric acid for >2 h; then, rinse with distilled water and dry in a clean environment.

Box 5.2 Calibration of Microwave Equipment (Cavity-Type Instrumentation)

Measurement of the available microwave power is assessed by measuring the temperature rise in 1 kg of water exposed to microwave radiation for a fixed period of time. By producing a multiple-point calibration, the analyst

Continued on page 63

■ *Continued from page 62* ■

can relate power (in watts) to the partial power setting of the microwave system. Typically for a 600 W microwave system, the following power settings are measured: 100, 99, 98, 97, 95, 90, 80, 70, 60, 50 and 40%. This is carried out as follows.

Equilibriate a large volume of water to room temperature ($23 \pm 2°C$). Then, weigh 1 kg of reagent ($1000.0 \text{ g} \pm 0.1$ g) into a fluorocarbon beaker (or similar container that does not absorb microwave energy, e.g. *not* glass). The initial temperature of the water should be $23 \pm 2°C$, measured to $\pm 0.05°C$. The covered beaker is then subjected to microwave radiation at the desired partial power settings for 2 min under normal conditions of operation, i.e. exhaust fan on. After the elapsed time, immediately remove the beaker, add a magnetic stirring bar and stir vigorously. Record the maximum temperature within the first 30 s to $\pm 0.05°C$. By using a new water sample for each subsequent measurement, determine three measurements at each power setting. The apparent power absorbed by the sample (P), measured in watts ($W = J \text{ s}^{-1}$), can then be determined as follows:

$$P = \frac{K \times C_p \times m \times \Delta T}{t}$$

where K is the factor for converting thermochemical calories per second to watts ($= 4.184$), C_p the heat capacity (at constant pressure), thermal capacity or specific heat ($\text{cal g}^{-1} °C^{-1}$) of water, m the mass of water sample (g), ΔT the final temperature minus the initial temperature ($°C$), and t the time (s).

Using the experimental conditions of 2 min and 1 kg of distilled water (heat capacity at $25°C$ is $0.9997 \text{ cal g}^{-1} °C^{-1}$), the calibration equation simplifies to the following:

$$P = \Delta T \times 34.86$$

The heating of acids singly or in combination, under pressure, in microwave ovens requires a level of caution to be exercised in the use of such systems. In general then, you should:

- Always consult the manufacturer's safety and operational recommendations prior to use.
- Always check the condition of the digestion vessel liners, and (specifically) the outer layers for acid attack and wear.
- Note that the use of domestic microwave ovens is not advisable for this type of work; they are neither acid resistant or have suitable safety features for scientific work.

The recommendations for employing microwave digestion are as follows:

- Laboratory coats, safety glasses and protective gloves should be worn.

- Only use a single acid mixture and volume with a batch in the microwave oven. This should ensure consistent reaction conditions throughout all vessels.

- For samples of known volatility or containing easily oxidizable organic material, reduce the sample mass to 0.10 g and observe the reaction prior to capping the vessels. If no reaction occurs, the sample weight may be increased to 0.25 g. All samples known or suspected to contain between 5–10% organic material should be pre-digested in a fume cupboard for at least 15 min.

- Always open and close digestion vessels in a fume cupboard due to the toxic and corrosive nature of concentrated acid fumes and their reactive products.

- In the case of hydrofluoric acid, extra caution is needed (see Table 5.1). In addition, as HF can attack glassware only plasticware should be used, e.g. volumetric flasks and beakers. The addition of boric acid, after sample digestion, to complex the fluoride will allow samples to be analysed with standard apparatus, as in the case of inductively coupled plasma (ICP) methods (see Chapter 11).

DQ 5.3

You have been asked to analyse a contaminated land site for a range of metals by employing an atomic spectroscopic technique that requires you to convert the soil samples obtained into aqueous form using concentrated acid. Which approach would you use?

Answer

In your initial considerations, you would need to think about the following questions. How many samples are required to be analysed? How much sample do you have for decomposition? Do you have any time constraints? Can you analyse the samples immediately (and hence require fast decomposition) or is this not an issue? Based on the answers to these questions, you could then decide between acid decomposition using either conventional or microwave heating. Obviously, the answer may rest with whether you have the choice of both methods of heating or not. In the latter case, you probably would not go with the additional expense of acquiring the other source of heating.

5.4.3 Fusion

Some substances, e.g. silicates and oxides, are not normally destroyed by the action of acid. In this situation, an alternative approach is required. Fusion involves the addition of an excess (10-fold) of reagent to the sample (finely ground) which is placed in a metal crucible, e.g. Pt, followed by heating in a muffle furnace (300–1000°C). After heating for a period of time (minutes to hours), a clear 'melt' should result, thus indicating completeness of the decomposition process. After cooling, the melt will dissolve in a mineral acid. Typical reagents include sodium carbonate (12–15 g of flux required per g of sample; heat to 800°C; dissolve with HCl), lithium meta- or tetraborate (10–20 fold excess of flux required; heat to 900–1000°C; dissolve with HF), and potassium pyrosulfate (10–20 fold excess of flux required; heat to 900°C; dissolve with H_2SO_4). The obvious addition of excess reagent (flux) can lead to a high risk of contamination. In addition, the high salt content of the final solution may lead to problems in the subsequent analysis.

DQ 5.4
What problems might result from a high sample salt content?

Answer

A high salt content can cause problems in the analysis step. For example, a high salt content can block the nebulizer used for sample introduction in both flame atomic absorption spectroscopy and inductively coupled plasma-based techniques (see Chapter 11).

5.5 Speciation Studies

Speciation is defined as 'the process of identifying and quantifying the different defined species, forms or phases present in a material' or 'the description of the amounts and types of these species, forms or phases present'. In some cases, it is possible to identify, by using single or sequential extractions, operationally defined determinations which identify 'groups' of metals without clear identification. In this situation, it is possible to refer to, for example, ethylenediaminetetraacetic acid (EDTA)-extractable trace metals. The reasons why speciation is important is that metals and metalloids can be present in many forms, some of which are toxic.

DQ 5.5
Are you aware of any chemical forms in which metals and metalloids can be present?

Answer

As an example, chromium is found in two different oxidation states, i.e. Cr(III) and Cr(VI). However, while Cr(III) is beneficial to humans (within a certain concentration range), Cr(VI) is toxic, thus leading to the formation of cancers. This particular information has recently been brought to the attention of the public via the film 'Erin Brockovich', starring Julia Roberts. This film, which was released in 2000, is a true story based on the discharge into groundwater of Cr(VI) from the Pacific Gas and Electric (PG and E) Company in the USA and the subsequent legal action for damages.

Information related to species-specific issues is rather limited, however, with details currently only known for a limited number of metals and metalloids, e.g. arsenic, tin and lead. While the example for Cr was obviously water-related, these issues are also important in soil (and sediment analysis), as well as for food matrices. The latter, for obvious reasons, inasmuch as we may well be eating the food, while in the case of the former because food crops may be grown in contaminated soil prior to consumption and entry into the food chain.

One approach to investigate the speciation of metals and metalloids in environmental samples has been the linking of chromatographic separation with quantitation by atomic spectroscopy. In this situation, the use of a suitable chromatographic technique, e.g. gas or liquid chromatography, is being used to separate a metal complex prior to detection of the metal by an atomic spectroscopic technique, e.g. inductively coupled plasma–mass spectrometry (see Chapter 11). However, while chromatography is capable of highly reproducible separations, it can only accept liquid samples. In this situation, the use of specific methods of extraction are required to remove the extract, without altering its chemical form (speciation) from the sample matrix.

5.6 Selected Examples of Metal Speciation

5.6.1 Mercury

All forms of mercury are considered to be poisonous. However, it is methylmercury (or as the chloride, CH_3HgCl) which is considered to be the most toxic because of its ability to bioaccumulate in fish. The best example of the toxicity of methylmercury occurred in Minamata in Japan in 1955. It was found that methylmercury-contaminated fish consumed by pregnant women resulted in the new-born children having severe brain damage (Minamata disease). As a consequence of these initial findings, methylmercury is routinely monitored for in fish.

A variety of methods exist for the extraction, clean-up and subsequent analysis of methylmercury in samples. Most methods involve a solvent extraction, often with toluene, of the methylmercury from the sample, followed by separation by gas chromatography with an electron-capture detector. The specific details for methylmercury extraction from fish tissue are shown in Figure 5.10. A similar approach can also be applied for the extraction and analysis of methylmercury from sediment. A summary of the methods used for the determination of methylmercury in fish and sediment matrices is given in Table 5.2. This has also been recently reviewed [13].

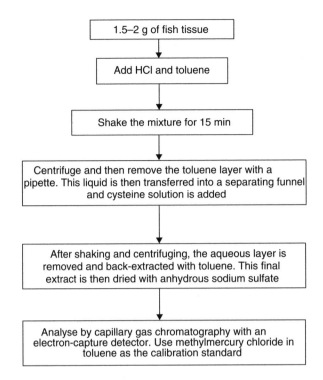

Figure 5.10 A typical procedure for the extraction of methylmercury [11].

Table 5.2 Summary of the various methods used for methylmercury determination [12]

Extraction technique[a]	Separation technique[b]	Detection[b]
Fish tissue		
250 mg of sample extracted twice with HCl, followed by UV irradiation	Ion-exchange (Dowex 1×8 100–200 mesh)	AAS
1500–2000 mg of pre-wetted sample extracted with HCl and toluene, followed by LLE with cysteine and back-extraction into toluene[c]	cGC	ECD
800–1000 mg of sample treated with H_2SO_4, followed by toluene extraction	cGC	ECD
400–500 mg of sample treated with HCl and extracted into toluene and cysteine, followed by back-extraction into toluene	cGC	ECD
500 mg of sample treated with NaCl/HCl and extracted into benzene and re-extracted with a thiosulfate solution	cGC	ECD
50 mg of sample extracted with H_2SO_4/MeCOOH into water	cGC	AFS
500 mg of sample treated with NaCl/HCl, extracted into benzene and re-extracted with a thiosulfate solution	—	ETAAS

(continued overleaf)

Table 5.2 (*continued*)

Extraction technique[a]	Separation technique[b]	Detection[b]
100 mg of sample treated with hot HCl and extracted into toluene. Extracts treated with thiosulfate solution and heated. Mercury species converted to hydride with NaBH$_4$	cGC	FTIR
50 mg of sample treated with NaCl, followed by HCl, and extracted with DDTC (pH 9) into toluene. Mercury species butylated with Grignard reagent	cGC	MIP
300 mg of sample distilled with H$_2$SO$_4$ in 20% KCl at 145°C. Mercury species derivatized with 1% NaBEt$_4$ in acetic acid	GC	CVAFS
Sediment		
2000 mg of sample treated with HCl, and extracted into toluene and cysteine and back-extracted into toluene	cGC	ECD
250 mg of sample treated with HCl, followed by toluene extraction	cGC	ECD
200 mg of sample treated with H$_2$SO$_4$/NaCl and extracted into toluene and thiosulfate solution. Mercury species converted to hydride with NaBH$_4$	cGC	CVAAS
250 mg of sample treated with HCl and extracted into toluene, followed by clean-up with cysteine solution, and back-extraction into toluene	cGC	CVAAS
500 mg of sample microwave-digested with HNO$_3$, followed by derivatization with NaBEt$_4$	cGC	QFAAS
200 mg of pre-wetted sample distilled with H$_2$SO$_4$/KCl at 145°C. Mercury species derivatized with NaBEt$_4$ in acetate buffer	GC	CVAFS
25 mg of sample distilled with H$_2$SO$_4$/NaCl/H$_2$O. Mercury species derivatized with NaBEt$_4$	GC	CVAAS
250 mg of sample distilled with H$_2$SO$_4$/NaCl/H$_2$O at 140°C. Distillate treated with 5% NH$_4$OAc at pH 6. Mercury species complexed with 0.5% SPDC	HPLC	CVAAS
200 mg of sample distilled with H$_2$SO$_4$/NaCl at 145°C. Distillate treated with NH$_4$OAc at pH 6. Mercury species complexed with 0.5 mmol l^{-1} SPDC, followed by on-line UV irradiation, and reduction by NaBH$_4$	HPLC	ICP–MS
1000 mg of sample treated with HCl, extracted into toluene and back-extracted into Na$_2$S$_2$O$_3$. Complexation by mercaptoethanol performed on-line	HPLC followed by on-line oxidation (H$_2$SO$_4$ and CuSO$_4$) and reduction with SnCl$_2$	CVAAS
500 mg of sample extracted with supercritical CO$_2$ and eluted with toluene. Mercury species butylated with Grignard reagent	cGC	MIP

[a]UV, ultraviolet; LLE, liquid–liquid extraction; DDTC, diethyldithiocarbamate; SPDC, sodium pyrrolidinedithiocarbamate.
[b]GC, gas chromatography; cGC, capillary gas chromatography; HPLC, high performance liquid chromatography; AAS, atomic absorption spectroscopy; ECD, electron-capture detection; AFS, atomic fluorescence spectroscopy; ETAAS, electrothermal atomization atomic absorption spectroscopy; FTIR, Fourier-transform infrared (spectroscopy); MIP, microwave-induced plasma; CVAFS, cold-vapour atomic fluorescence spectroscopy; QFAAS, quartz-furnace atomic absorption spectroscopy; ICP–MS, inductively coupled plasma–mass spectrometry.
[c]See Figure 5.10 for a more complete description.

5.6.2 Tin

Butyl- and phenyltin compounds are known to be toxic in the marine environment [14]. Historically, tributyltin (TBT) has been released into the marine environment from the leaching of TBT-based antifouling paints used on the undersides of boats and ships. Triphenyltin (TPT) has also been used as an antifouling agent in paint and in herbicide formulations.

Methods have therefore been developed to extract tin compounds from sediment, particularly in estuarine environments. The specific details for organotin compound (excluding monoalkyl tin compounds) extraction from sediment, sewage sludge, weeds and fish tissue are shown in Figure 5.11, while the conditions employed for the analysis of organotin extracts by electrothermal (graphite furnace) atomic absorption spectroscopy are shown in Table 5.3. Calibration

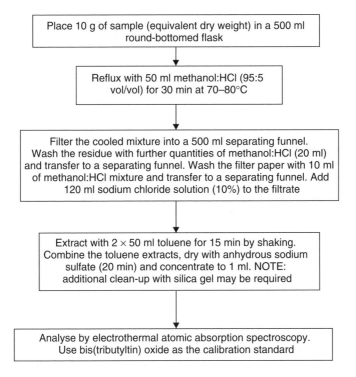

Figure 5.11 A typical procedure used for the extraction of organotins [15].

Table 5.3 Conditions employed for the analysis of organotin extracts using electrothermal atomic absorption spectroscopy [15]

Parameter	Setting/Condition
Hollow-cathode lamp current	6 mA
Wavelength	286.3 nm
Sample volume	5 μl
Background correction required	Yes
Graphite tube	Lanthanum coated
Graphite furnace programme (gas flow of 50 ml min^{-1} argon is on during Steps 1–4)	Step 1: 50°C for 10 s
	Step 2: 85°C for 10 s
	Step 3: 600°C for 5 s
	Step 4: 900°C for 5 s
	Step 5: 2700°C for 7 s
	(absorbance recorded)

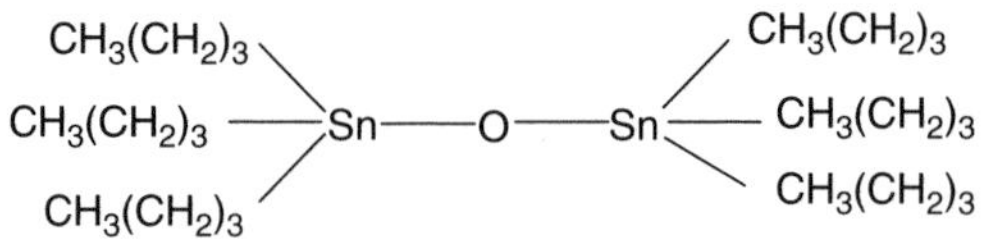

Figure 5.12 The molecular structure of bis(tributyltin) oxide (TBTO); the molecular weight of this compound is 596.12, while the atomic weight of tin is 118.69.

is carried out using bis(tributyltin) oxide (TBTO) (Figure 5.12), by accurately weighing 0.0502 g of TBTO, dissolving in glacial acetic acid and making up to a final volume of 200.00 ml with more acid. A search of the scientific literature will undoubtedly identify a range of procedures for the extraction and separation of organotin species. A summary of the methods used for the determination of organotin compounds in sediments is shown in Table 5.4. A recent review, however, highlights the determination of organotin compounds by using high performance liquid chromatography [17].

SAQ 5.3

By reference to Table 5.4, can you identify any similarities between the different extraction methods used for butyltin determination from sediments?

5.6.3 Arsenic

Arsenic occurs in many chemical forms in the environment, i.e. arsenite As(III), arsenate As(v), monomethylarsonic acid (MMAA), dimethylarsinic acid (DMAA), arsenobetaine (AsB) and arsenochloine (AsC) (Table 5.5), with a

Table 5.4 Summary of the various methods used for butyltin determination in sediments [16]

Extraction technique[a]	Separation technique[b]	Detection[c]
1 g of sample extracted with 20 ml acetic acid, followed by centrifugation at 4000 rpm for 15 min. Extract derivatized with 10% NaBH$_4$ in 1% NaOH	GC	QFAAS
2 g of sample extracted with 20 ml acetic acid and stirred for 12 h, followed by centrifugation at 4000 rpm for 15 min. Extract derivatized with 5% NaBH$_4$	GC	QFAAS
2 g of sample extracted with 10 ml acetic acid and mechanically stirred for 16 h, followed by centrifugation at 4000 rpm for 5 min. Extract derivatized with 4% NaBH$_4$	GC	QFAAS
2 g of sample, pre-wetted with 2 ml water, extracted with 8 ml of 0.1% NaOH in methanol and back-extracted into 2 ml hexane. Extract derivatized with NaBH$_4$, followed by back-extraction into hexane	cGC	FPD
1 g of sample extracted with 5 ml HCl and 10 ml toluene, with mechanical shaking for 15 h. Extract derivatized with 2% NaBEt$_4$.	cGC	FPD
3.5 g of sample extracted using SFE with CO$_2$ and 20% methanol, after HCl addition. Extract derivatized with Grignard reagent, using 2 mol l^{-1} EtMgCl in tetrahydofuran	cGC	FPD
5 g of sample, stirred for 1 h with 50 ml of a HBr–water mixture and then extracted for 2 h with tropolone/pentane, followed by centrifugation at 1000 rpm for 10 min. Organic phase was then dried with Na$_2$SO$_4$ and reduced in volume to 0.5 ml. Extract derivatized with Grignard reagent, using 1 mol l^{-1} pentylmagnesium chloride in diethyl ether, for 1 h. Then, excess Grignard reagent destroyed by dropwise addition of H$_2$SO$_4$. Organic layer extracted twice with 5 ml pentane; all extracts combined and dried with anhydrous Na$_2$SO$_4$. Resultant extract evaporated to 0.5 ml and purified by passing through a 'Florisil' column with pentane. The latter extract reduced in volume to 0.5 ml under N$_2$	cGC	FPD
2 g of sample ultrasonically extracted for 30 min with 4 ml deionized water, 2 ml acetic acid, 2 ml DDTC in pentane and 25 ml hexane. Hexane phase was collected and sediment was back-extracted with 25 ml hexane. Combined hexane extracts evaporated to dryness, and then re-dissolved with 250 ml n-octane. Extract derivatized with Grignard reagent, using 2 mol l^{-1} pentylmagnesium bromide in diethyl ether	cGC	QFAAS
1 g of sample ultrasonically extracted with 20 ml diethyl ether/HCl in tropolone. Extract derivatized with Grignard reagent, using 2 mol l^{-1} pentylmagnesium bromide in diethyl ether	cGC	MS
0.5 g of sample extracted with methanol/tropolone after addition of HCl. Extract derivatized with Grignard reagent, using 2 mol l^{-1} pentylmagnesium bromide in diethyl ether, followed by clean-up with silica gel	cGC	MS
2 g of sample extracted with acetic acid and back-extracted into toluene. Extract pre-concentrated by solvent evaporation	HPLC	ICP-MS
25 g of sample leached with 150 ml HCl (0.5 mol l^{-1}) for 24 h at room temperature. Slurry centrifuged at 1500 rpm for 30 min. Solid material washed with 0.5 mol l^{-1} HCl, followed by centrifugation. Combined supernatant filtered and then extracted with dichloromethane. Organic phase was added to 5 ml LiCl electrolyte (0.2 mol l^{-1}) in ethanol	—	Polarography

[a] SFE, supercritical fluid extraction; DDTC, diethyldithiocarbamate.
[b] GC, gas chromatography; cGC, capillary gas chromatography; HPLC, high performance liquid chromatography.
[c] QFAAS, quartz-furnace atomic absorption spectroscopy; FPD, flame photometric detection; MS, mass spectrometry; ICP–MS, inductively coupled plasma–mass spectrometry.

Table 5.5 Examples of arsenic compounds found in environmental samples

Compound	Formula
Arsenious acid; arsenite; As(III)[a]	$HAsO_2$
Arsenic acid; arsenate; As(v)[a]	H_3AsO_4
Monomethylarsonic acid (MMAA)[a]	$H_2(CH_3)AsO_3$
Dimethylarsinic acid (DMAA)[a]	$H(CH_3)_2AsO_2$
Arsenobetaine (AsB)	$(CH_3)_3As^+CH_2COOH$
Arsenocholine (AsC)	$(CH_3)_3As^+CH_2CH_2COOH$

[a]Compounds forming gaseous species: As(III) and As(v) form AsH_3; MMAA forms monomethylarsine, CH_3AsH_2; DMAA forms dimethylarsine, $(CH_3)_2AsH$.

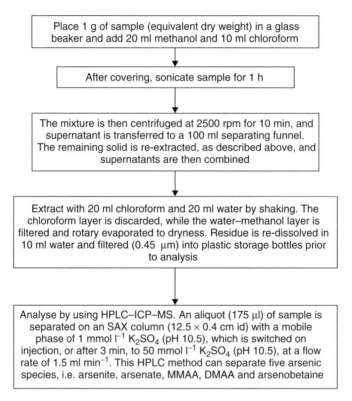

Figure 5.13 Procedure used for organoarsenical extraction – 'Method 1' [18]: HPLC–ICP–MS, high performance liquid chromatography–inductively coupled plasma–mass spectrometry; MMAA, monomethylarsonic acid; DMAA, dimethylarsinic acid. Note: original extraction procedure based on that of Beauchemin *et al.* [19].

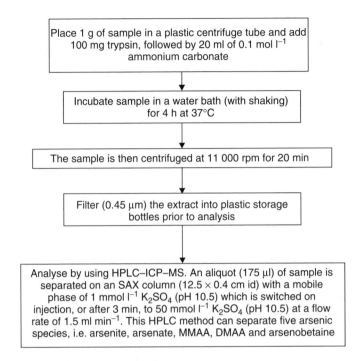

Figure 5.14 Procedure used for organoarsenical extraction – 'Method 2' [18]: HPLC–ICP–MS, high performance liquid chromatography–inductively coupled plasma–mass spectrometry; MMAA, monomethylarsonic acid; DMAA, dimethylarsinic acid. Note: original extraction procedure based on that of Crews *et al.* [20].

range of toxicities, e.g. As(III) is toxic. In fish tissue, the main form of arsenic is arsenobetaine, which is non-toxic. A variety of methods exist for the extraction, clean-up and subsequent analysis of arsenic species in samples. Two specific methods used for the extraction and subsequent analysis of arsenic species in fish tissue are shown in Figures 5.13 and 5.14. Often, the extracts are analysed by using combined chromatography separation with atomic spectroscopic detection. While a search of the scientific literature will identify a variety of approaches and procedures that have been applied, a recent review highlights the main separation techniques [21]. A summary of the methods used for the determination of organoarsenicals in fish tissue is given in Table 5.6.

SAQ 5.4

By reference to Table 5.6, can you identify any similarities between the different extraction methods used for organoarsenical determination from sediments?

Table 5.6 Summary of the various methods used for organoarsenical determination in fish tissue [22]

Extraction technique[a]	Separation technique[b]	Detection[c]
1 g of sample ultrasonically extracted with water/methanol (1:1 (vol/vol)). Extract evaporated, followed by SPE clean-up	Cation-exchange, followed by post-column hydride generation with $NaBH_4$ after UV irradiation	ICP–AAS
2 g of sample ultrasonically extracted with water/methanol (1:1 (vol/vol)). Extract evaporated, followed by silica column clean-up	Anion-exchange, followed by post-column hydride generation with $NaBH_4$ after UV irradiation	QFAAS
2 g of sample extracted with water	GC, followed by post-column hydride generation with $NaBH_4$ after UV irradiation	QFAAS
1 g of sample ultrasonically extracted with water/methanol (1:1 (vol/vol)). Extract evaporated, diluted with water and filtered	HPLC followed by post-column hydride generation with $NaBH_4$	ICP–AES
1 g of sample extracted with water/methanol (1:1 (vol/vol)) by stirring. Extract evaporated, diluted with water and filtered, followed by SPE	HPLC (anion-exchange), followed by post-column hydride generation with $NaBH_4$	QFAAS
1 g of sample ultrasonically extracted with water/methanol (1:1 (vol/vol)). Extract evaporated, diluted with water and filtered	HPLC (anion-exchange), followed by post-column hydride generation with $NaBH_4$	QFAAS
1 g of sample ultrasonically extracted with water/methanol (1:3 (vol/vol)). Extract evaporated, diluted with water and filtered (SPE)	HPLC (ion-pair)	ICP–MS
1 g of sample ultrasonically extracted with water/methanol (1:1 (vol/vol)). Extract evaporated, diluted with water and filtered	HPLC (anion-exchange)	ICP–MS
0.5 g of sample extracted using mechanical stirring with 0.1 g trypsin and ammonium bicarbonate (pH 8)	HPLC (anion-exchange)	ICP–MS
1 g of sample microwave-extracted with water. Extract filtered	HPLC (anion-exchange)	ICP–MS
150 mg of sample ultrasonically extracted with methanol/methane (5:2 (vol/vol)) and back-extracted into water. Extract evaporated	HPLC (caton-exchange)	ICP–MS

[a]SPE, solid-phase extraction.
[b]UV, ultraviolet; GC, gas chromatography; cGC, capillary gas chromatography.
[c]ICP–AAS, inductively coupled plasma–atomic absorption spectroscopy; QFAAS, quartz-furnace atomic absorption spectroscopy; ICP–AES, inductively coupled plasma–atomic emission spectroscopy; ICP–MS, inductively coupled plasma–mass spectrometry.

5.6.4 Chromium

Trivalent chromium (Cr(III)) is an essential element for man as it is involved in glucose, lipid and protein metabolism. In contrast, hexavalent chromium (Cr(VI)) is a potent carcinogen. It is therefore essential to be able to distinguish between these two different oxidation states of chromium. A variety of methods exist for the extraction and subsequent determination of Cr(VI). A specific method for the extraction of Cr(VI) from soluble, adsorbed and precipitated forms of chromium compounds in soils, sludges, sediments and similar materials is shown in Figure 5.15. Among the methods suitable for the analysis of Cr(VI) is a colorimetric method based on diphenylcarbazide (Figure 5.16). This method is suitable for the determination of Cr(VI) within the range $0.5-50$ mg l^{-1}. Interferences from molybdenum and mercury (up to 200 mg l^{-1} can be tolerated) are possible at low concentrations of Cr(VI); interference is also possible from vanadium (concentrations up to $10 \times$ Cr(VI) can be tolerated). A calibration curve is generated by serial dilution of a 50 mg l^{-1} Cr(VI) stock solution prepared from potassium dichromate (dissolve 141.4 mg of potassium dichromate ($K_2Cr_2O_7$) in reagent water and dilute to 1 l) and then following the same procedure as used for the sample (see Figure 5.16). A search of the scientific literature will undoubtedly identify a range of procedures for the extraction and separation of chromium species. A recent review, however, highlights the current methods used for the determination of chromium species in solution [25].

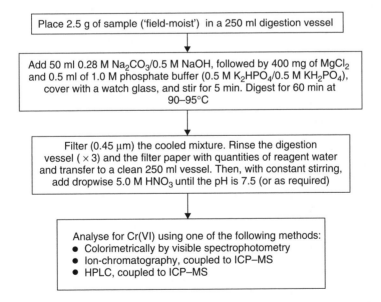

Figure 5.15 Procedure used for the extraction of hexavalent chromium [23]: ICP–MS, inductively coupled plasma–mass spectrometry.

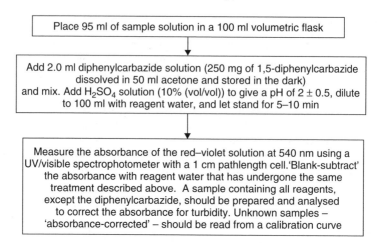

Figure 5.16 A colorimetric method, based on diphenylcarbazide, used for the analysis of hexavalent chromium[24].

5.7 Selective Extraction Methods

5.7.1 Plant Uptake Studies

Extraction procedures for plant uptake studies have existed for many years. These species are defined by their function, e.g. plant-available forms. The species defined in this way, however, are unlikely to be distinct chemical forms but may include a range of chemical entities that share the same function, e.g. are all available to plants. This type of functionality, using selective chemical extractants, has been widely employed in soil and agricultural laboratories to diagnose or predict toxicity deficiencies in crops and in animals eating such crops. Table 5.7 illustrates the diversity of this type of selective extraction. It is noted that the methods, evolved over many years on an empirical basis, are both element-specific and crop-specific.

5.7.2 Soil Pollution Studies

In the case of metal-polluted soils, a different approach for extraction has been identified. Extractants have been studied that allow isolation of a particular metal-containing soil phase, e.g. *exchangeable* materials. The information can then be used to identify the potential likelihood of metal release, transformation, mobility or availability as the soils are exposed to weathering, to pH changes, and to changes in land use, with the implications for environmental risk assessment that may result. As a result of the non-specific nature of most of these extraction methods, it is necessary to define limits in which the extractant will operate. Table 5.8 identifies a variety of reagents or methods of isolation for the removal

Table 5.7 Selective-extraction methods which are diagnostic of plant uptake [26]. Reprinted from *Sci. Total Environ.*, **178**, Ure, A. M., 'Single extraction schemes for soil analysis and related applications', 3–10, Copyright (1996), with permission from Elsevier Science

Extractant	Element	Correlated plant content
Water	Cd, Cu, Zn	Wheat, lettuce
EDTA[a] ($0.05 \ mol \, l^{-1}$)	Cd, Cu, Ni, Pb, Zn	Arable crops
EDTA[a] ($0.05 \ mol \, l^{-1}$)	Se, Mo	Greenhouse crops
DTPA[b]	Cd, Cu, Fe, Mn, Ni, Zn	Beans, lettuce, maize, sorghum, wheat
Acetic acid (2.5 vol%)	Cd, Co, Cr, Ni, Pb, Zn	Arable crops, herbage
Ammonium acetate, pH 7 ($1 \ mol \, l^{-1}$)	Mo, Ni, Pb, Zn	Herbage, oats, rice, sorghum, Swiss chard
Ammonium acetate:EDTA[a] ($0.5 \ mol \, l^{-1} : 0.02 \ mol \, l^{-1}$)	Cu, Fe, Mn, Zn	Wheat
$CaCl_2$ ($0.05 \ mol \, l^{-1}$)	Cd, Pb	Vegetable(s)
$NaNO_3$ ($0.1 \ mol \, l^{-1}$)	Cd, Pb	Vegetable(s)
Ammonium nitrate ($0.05 \ mol \, l^{-1}$)	Cd, Pb	Vegetable(s)

[a]EDTA, ethylenediaminetetraacetic acid (diammonium salt).
[b]DTPA, $0.005 \ mol \, l^{-1}$ diethylenetriaminepentaacetic acid + $0.1 \ mol \, l^{-1}$ triethanolamine + $0.01 \ mol \, l^{-1}$ $CaCl_2$.

Table 5.8 Various extraction procedures used to isolate nominal soil/sediment phases [26]. Reprinted from *Sci. Total Environ.*, **178**, Ure, A. M., 'Single extraction schemes for soil analysis and related applications', 3–10, Copyright (1996), with permission from Elsevier Science

Phase isolated or extracted	Reagent or method of isolation
Water-soluble, soil solution, sediment-pore water	Water, centrifugation, displacement, filtration, dialysis
Exchangeable	$MgCl_2$
	NH_4OAc
	HOAc
Organically bound	$Na_4P_2O_7$
	H_2O_2 (at pH 3)/NaOAC
Carbonate	NaOAc at pH 5 (HOAc)
Mn oxide-bound	$NH_2OH.HCl$
Fe (amorphous) oxide	$(NH_4)_2C_2O_4$ in dark
Fe (crystalline) oxide	$(NH_4)_2C_2O_4$ in UV light

of metals from defined soil phases. It can be seen that a range of different phases and approaches are possible and these are now briefly discussed in the following.

- *Water-soluble, soil solution, sediment-pore water.* This phase contains the most mobile and hence potentially available metal species.

- *Exchangeable species.* This phase contains weakly bound (electrostatically) metal species that can be released by ion-exchange with cations such as Ca^{2+}, Mg^{2+} or NH_4^+. Ammonium acetate is the preferred extractant as the complexing power of acetate prevents re-adsorption or precipitation of released metal ions. In addition, acetic acid dissolves the exchangeable species, as well as more tightly bound exchangeable forms.

- *Organically bound.* This phase contains metals bound to the humic material of soils. Sodium hypochlorite is used to oxidize the soil organic matter and release the bound metals. An alternative approach is to oxidize the organic matter with 30% hydrogen peroxide (acidified to pH 3), followed by extraction with ammonium acetate to prevent metal ion re-adsorption or precipitation.

- *Carbonate bound.* This phase contains metals that are dissolved by sodium acetate acidified to pH 5 with acetic acid.

- *Oxides of manganese and iron.* Acidified hydroxylamine hydrochloride releases metals from the manganese oxide phase with minimal attack on the iron oxide phases. Amorphous and crystalline forms of iron oxides can be discriminated between by extracting with acid ammonium oxalate in the dark and under UV light, respectively.

The diversity and complexity of the available approaches has identified the major difficulties associated with producing suitable guidelines that would allow comparisons between different laboratories and different countries in assessing metal mobility in the soil environment. This led to the development of single and sequential extraction procedures by the Standards, Materials and Testing (SM&T – formerly the (European) Community Bureau of Reference (BCR) Programme of the European Union (1987)). Single extractants evaluated included 0.05 mol l^{-1} EDTA, 0.43 mol l^{-1} acetic acid, and 1 mol l^{-1} ammonium acetate at pH 7.

DQ 5.6

Take a look at the results of this study in Table 5.9 below. As a result of these findings, the use of ammonium acetate was excluded as a suitable single extraction method. Why do you think this was?

Answer

Ammonium acetate was excluded due to its poor reproducibility, probably as a result of the low concentrations of metals extracted.

5.7.3 Single Extraction Procedures

The procedures for the extraction of metals using the two single extraction methods (0.05 mol l^{-1} EDTA and 0.43 mol l^{-1} acetic acid) are shown in Figures 5.17 and 5.18, respectively.

Table 5.9 Metal contents of soil, determined by using single extractants (mg kg^{-1} dry soil): interlaboratory results ($n = 35$) [27]

Metal	Cd	Cr	Cu	Ni	Pb	Zn
EDTA^a (0.05 mol l^{-1}) extracts						
Mean	23.1	8.05	162.0	16.3	255.0	492.0
RSD^b	9.2	25.9	8.5	13.0	11.8	5.8
Acetic acid (0.43 mol l^{-1}) extracts						
Mean	19.3	26.1	29.2	15.7	3.36	522.0
RSD^b	7.5	8.2	10.0	18.1	24.6	6.9
Ammonium acetate (1 mol l^{-1}) extracts						
Mean	3.43	1.39	5.65	1.42	2.21	18.4
RSD^b	10.9	40.6	23.4	22.5	26.8	22.5

^{a}Ethylenediaminetetraacetic acid.
^{b}*RSD*, relative standard deviation (%).

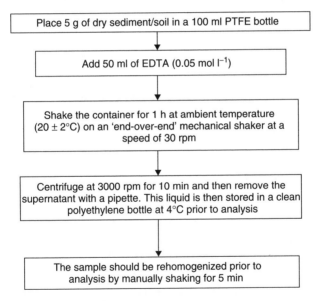

Figure 5.17 Procedure adopted in the single extraction method for metals (employing EDTA), as applied to the analysis of soils and sediments.

For these procedures (Figures 5.17 and 5.18), it is important to note the following:

- All laboratory ware should be made of borosilicate glass, polypropylene, polyethylene or PTFE, except for the centrifuge tubes which should be made of either borosilicate glass or PTFE.

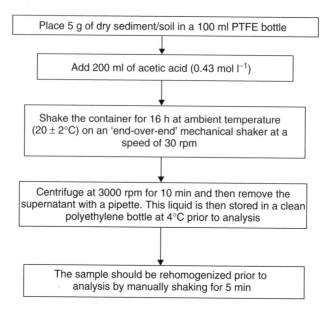

Figure 5.18 Procedure adopted in the single extraction method for metals (employing acetic acid), as applied to the analysis of soils and sediments.

- Clean all vessels in contact with samples or reagents with HNO_3 (4 mol l^{-1}) for at least 30 min, rinse with distilled water, clean with 0.05 mol l^{-1} EDTA and then rinse again with distilled water. Alternatively, clean all vessels in contact with samples or reagents with HNO_3 (4 mol l^{-1}) overnight and rinse with distilled water.

- Use a mechanical shaker, preferably of the 'end-over-end' type, at a speed of 30 rpm. Carry out the centrifugation at $3000 \times G$.

The reagents to be used are as follows:

- *Water.* Glass-distilled water is suitable. Analyse a sample of distilled water with each batch of extracts.

- *EDTA (0.05 mol l^{-1}).* Add in a fume cupboard, 146 ± 0.05 g of EDTA (free acid) to 800 ± 20 ml of distilled water and partially dissolve, by stirring, in 130 ± 5 ml of saturated ammonia solution (prepared by bubbling ammonia gas into distilled water). The addition of ammonia shall be continued until all of the EDTA has dissolved. The obtained solution should be filtered, if necessary, through a filter paper of porosity 1.4–2.0 μm, into a 10 l polyethylene bottle and diluted to 9.0 ± 0.5 l with distilled water. The pH should be adjusted to 7.00 ± 0.05 by the addition of a few drops of either ammonia or hydrochloric acid (as appropriate). The solution should then be made up to 10 l with distilled water to obtain an EDTA solution of 0.05 mol l^{-1}. Analyse a sample of each batch.

- *Acetic acid (0.43 mol l^{-1})*. Add, in a fume cupboard, 250 ± 2 ml of glacial acetic acid ('AnalaR' grade or similar) to about 5 l of distilled water in a 10 l polyethylene bottle and make up to 10 l with distilled water.

The single extraction procedure can be carried out for the following trace metals: Cd, Cr, Cu, Ni, Pb and Zn. A separate sub-sample of the sediment or soil should be dried in a layer of approximately 1 mm in depth in an oven at $105 \pm 2°$C for 2–3 h and then weighed. From this, a correction 'to dry mass' can be achieved and applied to all of the analytical results obtained (quantity per g dry sediment/soil).

While the complete experimental details for the single extraction methods are shown in Figures 5.17 and 5.18, it is important to note the following:

- The calibrating solutions should be made up with the appropriate extracting solutions.
- For every batch of extractions, a blank sample, i.e. a container with no sediment/soil, should be carried through the entire procedure.
- The sediment/soil should be in complete suspension during the extraction. If this is not the case, adjust the shaking speed to ensure that the suspension is maintained.

The results obtained from the certification of two certified reference materials using this approach are shown in Table 5.10.

DQ 5.7

Compare the data presented in Figure 2.3 above with those shown in Table 5.10.

Answer

In both cases, the results are presented as the 'certified values' with an 'uncertainty' or '$\pm$' value. This uncertainty is determined as one standard deviation of the mean value.

Finally, it is recommended that for inductively coupled plasma (ICP) analysis a final filtration (0.45 μm) is carried out in order to prevent nebulizer blockages. If graphite-furnace atomic absorption spectroscopy (GFAAS) is the method of analysis, it is recommended that the standard additions method of calibration is used (see Chapter 1).

An alternative approach using diethylenetriaminepentaacetic acid (DTPA) has also been evaluated (Figure 5.19). The DTPA solution is prepared as follows: dissolve 149.2 g triethanolamine (0.01 mol l^{-1}), 19.67 g DTPA (0.005 mol l^{-1})

Table 5.10 Extractable metal contents of two certified reference materials (CRM 483 and CRM 484) used in the analysis of soils or sediments [28]

Certified reference material/extract	Certified value (mg kg^{-1})	Uncertainty (mg kg^{-1})
CRM 483 (Sewage sludge-amended soil)		
EDTA extracts		
Cd	24.3	1.3
Cr	28.6	2.6
Cu	215.0	11.0
Ni	28.7	1.7
Pb	229.0	8.0
Zn	612.0	19.0
Acetic acid extracts		
Cd	18.3	0.6
Cr	18.7	1.0
Cu	33.5	1.6
Ni	25.8	1.0
Pb	2.10	0.25
Zn	620.0	24.0
CRM 484 (terra rossa soil)		
EDTA extracts		
Cd	0.51	0.03
Cu	88.1	3.8
Ni	1.39	0.11
Pb	47.9	2.6
Zn	152.0	7.0
Acetic acid extracts		
Cd	0.48	0.04
Cu	33.9	1.4
Ni	1.69	0.15
Pb	1.17	0.16
Zn	193.0	7.0
CRM 484 (terra rossa soil)		
EDTA extracts		
Cd	2.68	0.09
Cr	0.205	0.022
Cu	57.3	2.5
Ni	4.52	0.25
Pb	59.7	1.8
Zn	383.0	12.0
Acetic acid extracts		
Cd	1.34	0.04
Cr	0.014	0.003
Cu	32.3	1.0
Ni	3.31	0.13
Pb	15.0	0.5
Zn	142.0	6.0

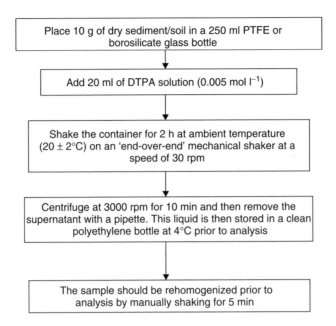

Figure 5.19 Procedure adopted in the single extraction method for metals (employing DTPA), as applied to the analysis of soils and sediments.

and 14.7 g calcium chloride in approximately 200 ml of distilled water. Allow the DTPA to dissolve and then dilute to 9 l. Adjust the pH to 7.3 ± 0.5 with HCl while stirring and dilute to 10 l. This working solution should be stable for several months.

5.7.4 Sequential Extraction Procedure

The sequential extraction procedure consists of three (main) stages, plus a final (residual fraction) stage (Figure 5.20), as follows:

- *Step 1.* Metals extracted during this step are those which are exchangeable and in the acid-soluble fraction. These includes weakly absorbed metals retained on the sediment surface by relatively weak electrostatic interaction, metals that can be released by ion-exchange processes and metals that can be co-precipitated with the carbonates present in many sediments. Changes in the ionic composition, influencing adsorption–desorption reactions, or lowering of pH, could cause mobilization of metals from such fractions.

- *Step 2.* Metals bound to iron/manganese oxides are unstable under reducing conditions. Changes in the redox potential (E_h) could induce the dissolution of these oxides and could also release adsorbed trace metals.

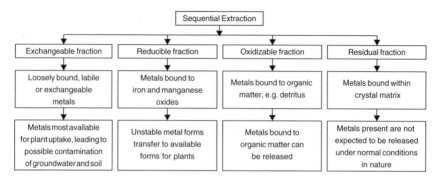

Figure 5.20 Overview of the sequential extraction method for metals, as applied to the analysis of soils and sediments.

- *Step 3.* Degradation of organic matter under oxidizing conditions can lead to a release of soluble trace metals bound to this component. Amounts of trace metals bound to sulfides might be extracted during this step.

It is common to analyse for trace metals in the residual fraction. In this situation, the latter should contain naturally occurring minerals which may hold trace metals within their crystalline matrix. Such metals are not likely to be released under normal environmental conditions. The residual fraction is digested by using a 'pseudo-total' approach with aqua regia as most metal pollutants are not silicate-bound. However, for complete digestion, hydrofluoric acid is required.

DQ 5.8

Why is the term 'pseudo-total' used?

Answer

Aqua regia is a good acid mixture for the digestion of soil and sediment samples. However, it cannot liberate from the matrix metal pollutants that are silicate-bound, i.e. part of the silicate 'backbone'. In this situation, if complete digestion is required then hydrofluoric acid must be used. As it is unlikely that the silicate-bound metals will leach from the soil or sediment, the use of aqua regia to give a 'pseudo-total' analysis is perfectly acceptable in this situation.

For the sequential extraction procedure (see details given in Figures 5.21–5.23), it is important to note the following:

- All laboratory ware should be made of borosilicate glass, polypropylene, polyethylene or PTFE, except for the centrifuge tubes which should be made of either borosilicate glass or PTFE.

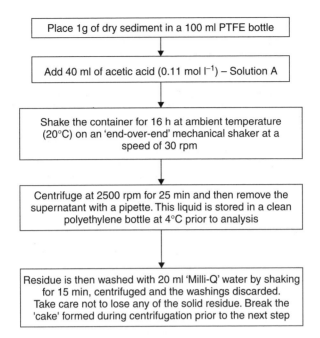

Figure 5.21 Details of Step 1 of the sequential extraction method (cf. Figure 5.20).

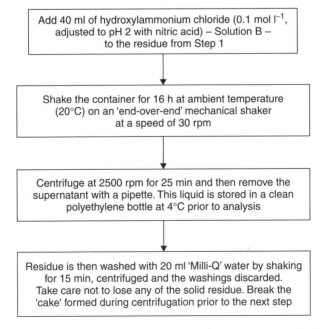

Figure 5.22 Details of Step 2 of the sequential extraction method (cf. Figure 5.20).

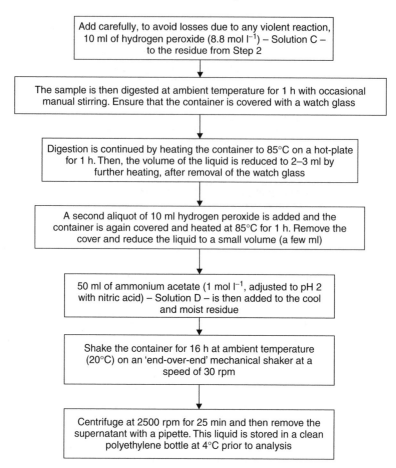

Figure 5.23 Details of Step 3 of the sequential extraction method (cf. Figure 5.20).

- Clean all vessels in contact with samples or reagents with HNO_3 (4 mol l^{-1}) overnight and then rinse with distilled water.
- Determine the blank as follows. To one vessel from each batch, taken through the cleaning procedure, add 40 ml of acetic acid (Solution A, see below). Analyse this blank solution along with the sample solution from Step 1 (as described in Figure 5.21).
- Use a mechanical shaker, preferably of the horizontal rotary or the 'end-over-end' type, at a speed of 30 rpm. Carry out the centrifugation at $1500 \times G$.

The reagents to be used for the three steps are as follows:

- *Water*. Glass-distilled water is suitable. Analyse a sample of distilled water with each batch of the Step 1 extracts.

- *Solution A (acetic acid; 0.11 mol l^{-1}).* Add in a fume cupboard, 25 ± 0.2 ml of glacial acetic acid ('AnalaR' grade or similar) to about 0.5 l of distilled water in a 1 l polyethylene bottle and make up to 1 l with distilled water. Then make up 250 ml of this solution (acetic acid; 0.43 mol l^{-1}) with distilled water to 1 l to obtain an acetic acid solution of 0.11 mol l^{-1}. Analyse a sample of each batch of Solution A.

- *Solution B (hydroxylamine hydrochloride or hydroxyammonium chloride; 0.1 mol l^{-1}).* Dissolve 6.95 g of hydroxylamine hydrochloride in 900 ml of distilled water. Acidify with HNO_3 to pH 2 and make up to 1 l with distilled water. Prepare this solution on the same day as the extraction is carried out. Analyse a sample of each batch of Solution B.

- *Solution C (hydrogen peroxide; 300 mg g^{-1}, i.e. 8.8 mol l^{-1}).* Use the H_2O_2 as supplied by the manufacturer, i.e. acid-stabilized to pH 2–3. Analyse a sample of Solution C.

- *Solution D (ammonium acetate; 1 mol l^{-1}).* Dissolve 77.08 g of ammonium acetate in 900 ml of distilled water, adjust to pH 2 with HNO_3 and make up to 1 l with distilled water. Analyse a sample of each batch of Solution D.

The sequential extraction procedure can be carried out for the following trace metals: Cd, Cr, Cu, Ni, Pb and Zn. A separate sub-sample of the sediment or soil should be dried in a layer of approximately 1 mm in depth in an oven at $105 \pm 2°$ C for 2–3 h and then weighed. From this, a correction 'to dry mass' can be obtained and applied to all of the analytical results obtained (quantity per g dry sediment/soil).

While the complete experimental details for the three steps in this sequential extraction method are shown in Figures 5.21–5.23, it is important to note the following:

- The calibrating solutions should be made up with the appropriate extracting solutions.

- For every batch of extractions, a blank sample, i.e. a container with no sediment/soil, should be carried through the entire procedure.

- The sediment/soil should be in complete suspension during the extraction. If this is not the case, adjust the shaking speed to ensure that the suspension is maintained.

The results obtained from the sequential extraction of a certified reference material (CRM 601) are shown in Table 5.11. Finally, it is recommended that for ICP analysis a final filtration (0.45 μm) is carried out in order to prevent

Table 5.11 The results obtained from the sequential extraction of a certified reference material (CRM 601) [28]

Metal	Certified value (mg kg^{-1})	Uncertainty (mg kg^{-1})
First step		
Cd	4.14	0.23
Cr	0.36	0.04
Cu	8.32	0.46
Ni	8.01	0.73
Pb	2.68	0.35
Zn	264.0	5.0
Second step		
Cd	3.08	0.17
Ni	6.05	1.09
Pb	33.1	10.0
Zn	182.0	11.0
Third step		
Cd	1.83	0.20
Ni	8.55	1.04
Pb	109.0	13.0

nebulizer blockages. If GFAAS is the method of analysis, it is recommended that the standard additions method of calibration is used (see Chapter 1).

5.7.5 Food Studies

In related studies, the simulated gastro-intestinal digestion of foodstuffs has been used as an indication of the bioavailable forms of metal species, and hence the potential for subsequent uptake from the intestine. A two-stage enzymolysis procedure has been developed and is shown in Figure 5.24. This procedure was developed to simulate *in vitro* enzyme treatments for investigating the speciation of metals from ingested foodstuffs. It consists of two stages, following by a post-enzymolysis acidification step. The first stage involves treatment of the foodstuff with pepsin at pH 2.5, followed in the second stage by pancreatin and amylase at pH 7. All enzymolysis extracts were analysed to assess the effectiveness of each stage to solubilize trace metals. Some selected results for the solubilization of trace metals from foodstuffs are shown in Table 5.12. The following conclusions were made: the solubility differs for different food items, thus reflecting the differences in the complex chemical nature of the original foodstuff; the enzymolysis itself brings about speciation changes, e.g. the reduction in solubility of cadmium from wholemeal bread and copper from tomato and spinach, after treatment with pancreatin and amylase; the pH affects the analyte solubility, with

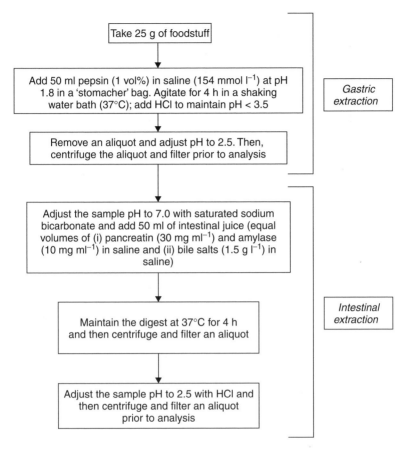

Figure 5.24 The gastro-intestinal extraction procedure used to investigate the speciation of metals from ingested foodstuffs [20].

acid-soluble levels being generally higher than those at neutral pH; the processing or preparation of food of similar origin may alter the solubility (not shown in Table 5.12).

An alternative approach for determining the bioaccessibility of metals from foodstuffs (including soil) has been developed. This approach, termed the physiologically based extraction test (PBET) [29, 30], has been applied to study the dietary levels of metals in Uganda, Africa [31]. It is postulated that endomyocardial fibrosis, a common cardiac disease in Africa, may be related to elevated levels of dietary cerium, and deficient levels of dietary magnesium. In Uganda, variation in dietary exposure to both Ce and Mg is probable due to variations in soil geochemistry and dietary habits, e.g. *geophagia*, the deliberate consumption of soil. In this instance, the soil was from termite nest soils and traditional

Table 5.12 Some selected results for the solubilization of trace metals in foodstuffs obtained by using an enzymolysis procedure [20]

Foodstuff	Metal															
	Cd				Zn				Fe				Cu			
	Total of fresh weight (mg kg^{-1})	Soluble portion (%)[a]			Total of fresh weight (mg kg^{-1})	Soluble portion (%)[a]			Total of fresh weight (mg kg^{-1})	Soluble portion (%)[a]			Total of fresh weight (mg kg^{-1})	Soluble portion (%)		
		G[b]	G+I[c]	G+I[d]		G[a]	G+I[c]	G+I[d]		G[a]	G+I[b]	G+I[c]		G[a]	G+I[b]	G+I[c]
Pig kidney	0.12	70	110	130	28	90	20	90	60	10	45	50	6.3	35	65	55
Crabmeat	1.26	120	50	120	83	90	25	95	64	25	5	45	21.0	75	80	85
100% beef burger	—	—	—	—	65	100	70	90	33	40	70	60	0.9	0	65	45
Wholemeal bread	0.04	100	50	50	20	105	0	95	29	30	35	55	2.9	65	90	55
Spinach	0.16	30	10	50	10	55	40	45	18	65	50	65	1.8	45	20	35
Tomato	—	—	—	—	2	90	100	75	6	60	80	55	0.5	80	55	50

[a] Levels quoted to nearest 10% (Cd) and nearest 5% (Zn, Fe, Cu).
[b] Gastric enzymes only, at pH 2.5.
[c] Gastric and intestinal enzymes, at pH 7.2–7.4.
[d] Gastric and intestinal enzymes, after subsequent adjustment of pH to 2.5.

herb-soil remedies. The details of this physiologically based extraction test are shown in Figure 5.25. Samples were analysed by using either inductively coupled plasma–atomic emission spectroscopy (ICP–AES) or inductively coupled plasma–mass spectrometry (ICP–MS) at Stages 1–3 (inclusive).

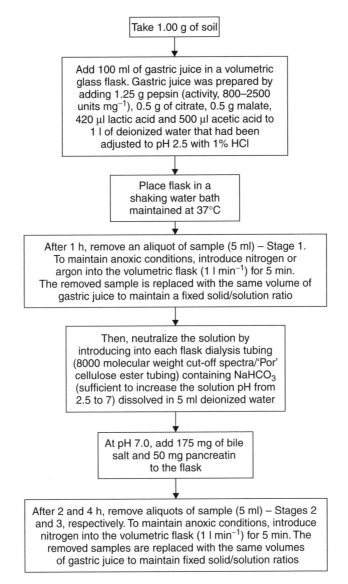

Figure 5.25 The physiologically based extraction test (PBET) used to determine the bioaccessibility of metals from foodstuffs [30].

5.8 Case Studies on Total and Selective Methods of Metal Analysis

5.8.1 Example 5.1: Total Metal Analysis of Soil, followed by Flame Atomic Absorption Spectroscopy

5.8.1.1 Digestion Conditions

The US EPA Method 3050B and its optional method (see Figures 5.2 and 5.3, respectively, above) were used, plus an aqua regia digest method (Figure 5.26). In each case, an accurately weighed 1.0000 g sample was used.

Comments Samples and sample blanks were filtered, cooled and made up to 100 ml with distilled water. Further dilutions were made, as appropriate. All glassware was soaked overnight in 10% nitric acid and then rinsed at least three times with distilled water.

5.8.1.2 Analysis by Flame Atomic Absorption Spectroscopy

The standards and samples ($n = 6$) were analysed by using a Perkin Elmer 100 FAAS system with an air–acetylene flame. The metals were determined at the following wavelengths: Cu, 324.8 nm; Fe, 248.3 nm; Pb, 217.0 nm; Mn, 279.5 nm; Ni, 232.0 nm; Zn, 213.9 nm. Calibrations were produced by serial dilution of 1000 μg ml^{-1} stock solutions in the range 0–10 μg ml^{-1}. All calibration graphs exhibited linear relationships for each metal, except zinc, i.e. $y = mx + c$. For zinc, a curved relationship was obtained, i.e. $y = ax^2 + bx + c$. All correlation coefficients were >0.96.

5.8.1.3 Typical Results

These are shown in Figure 5.27 [32].

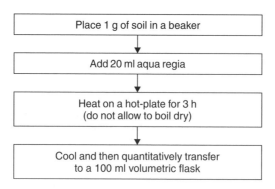

Figure 5.26 Procedure for the acid digestion of soils using aqua regia.

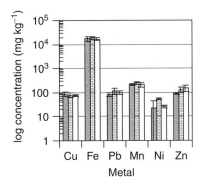

Figure 5.27 Typical results obtained for the total metal analysis of soil using flame atomic absorption spectroscopy. Digestion procedures: ■, aqua regia; ▩, US EPA Method 3050B; □, US EPA Method 3050B (optional) [32] (cf. DQ 5.9).

DQ 5.9

Comment on the results obtained in this study (see Figure 5.27).

Answer

It can be seen that all three digestion procedures have produced similar totals for each metal being analysed. Therefore, it can probably be concluded that on the basis of these results all three digestion procedures could be used for future studies.

5.8.2 Example 5.2: Total Metal Analysis of Soil Using X-Ray Fluorescence Spectroscopy – Comparison with Acid Digestion (Method 3050B), followed by Flame Atomic Absorption Spectroscopy

5.8.2.1 Digestion Conditions

No digestion procedures are necessary when using this analytical method.

5.8.2.2 Analysis by X-Ray Fluorescence Spectroscopy

The standards and samples were analysed by using a Spectro X-Lab 2000 X-ray fluorescence spectrometer operating with a Pd tube. The operating voltages ranged from 15 to 55 kV. The standards and samples of soil (4 g) were prepared as pressed pellets (pressure, 10 tonne) in the ratio 1:4 (wt/wt) of polymer (HWC, Hochst Wax C):soil. Calibration of standards and samples was achieved by using the 'Geology Analytical Program'.

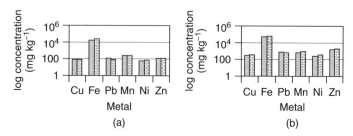

Figure 5.28 Comparison of the methods used in the analysis of (a) 'Soil A', and (b) 'Soil CONTEST 32.3A': ■, flame atomic absorption spectroscopy; ▦ X-ray fluorescence spectroscopy [32] (cf. DQ 5.10).

5.8.2.3 Typical Results

These are shown in Figure 5.28 [32].

DQ 5.10

Comment on the results obtained in this study (see Figure 5.28).

Answer

It can be seen that for both soil samples the XRF results obtained produce similar concentration values to those obtained after acid digestion using the US EPA Method 3050B, followed by FAAS. The use of XRF spectroscopy, which analyses the soil directly, provides faster analysis than that achieved when using the acid digestion/FAAS procedure.

5.8.3 Example 5.3: Sequential Metal Analysis of Soils, followed by Flame Atomic Absorption Spectroscopy

5.8.3.1 Sequential Extraction Conditions

In each case, an accurately weighed 1.0000 g sample was used. The sequential extraction method has been described in Figures 5.21–5.23.

Comments Samples and sample blanks were filtered, cooled and made up to 100 ml with distilled water. Further dilutions were made, as appropriate. All glassware was soaked overnight in 10% nitric acid and then rinsed at least three times with distilled water.

5.8.3.2 Analysis by Flame Atomic Absorption Spectroscopy

The standards and samples were analysed by using a Perkin Elmer 100 FAAS system with an air–acetylene flame. The metals were determined at the following wavelengths: Cu, 324.8 nm; Fe, 248.3 nm; Pb, 217.0 nm; Mn, 279.5 nm; Ni, 232.0 nm; Zn, 213.9 nm. Calibrations were produced by serial dilution of

1000 $\mu g\, ml^{-1}$ stock solutions in the range $0-10\ \mu g\, ml^{-1}$. All calibration graphs exhibited linear relationships for each metal, except zinc, i.e. $y = mx + c$. For zinc, a curved relationship was obtained, i.e. $y = ax^2 + bx + c$. All correlation coefficients were >0.96.

5.8.3.3 Typical Results

These are shown in Figure 5.29 [32].

DQ 5.11

Comment on the results obtained in this study (see Figure 5.29).

Answer

*It can be seen that the two soils behaved differently to the sequential extraction process. On average (for both soils) ca. 50% of the metal content was in the residual fraction, i.e. unlikely to be released. The majority of iron was **always** in the residual fraction (54% for 'Soil A' and 92% for the 'CONTEST' soil).*

The metals most available for plant uptake (Step 1 – see Figure 5.20 above) are as follows:

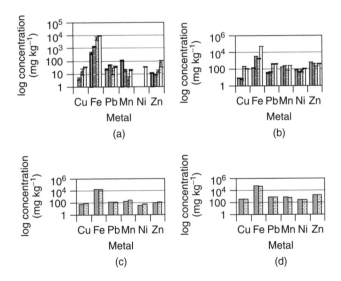

Figure 5.29 Sequential extraction results obtained for (a) 'Soil A', and (b) 'Soil CONTEST 32.3A': ■, Step 1; ■, Step 2; ▥, Step 3; □, residual fraction. Comparisons of the sequential extraction/flame atomic absorption spectroscopy (FAAS) total metal analysis and acid digestion/FAAS approaches for (c) 'Soil A', and (d) 'Soil CONTEST 32.3A': ■, sequential total; ▥, FAAS total [32] (cf. DQ 5.11).

- *Soil A: 72% Mn; 18% Pb; 10% Zn; 2% Fe (0% for Cu and Ni).*
- *Soil CONTEST 32.3A: 23% Mn; 4% Pb; 38% Zn and 0% Fe (2% for Cu and 28% for Ni).*

This information then needs to be related to the risk associated with the 'available' metals.

Summary

A variety of methods for the determination of total metals and metal species/fractions have been described. While the most common method of sample decomposition of environmental samples is probably acid digestion, other approaches, most notably the use of fusion, have some specialist application areas, e.g. geological samples. The area of metal speciation is an ever expanding one, with new approaches being developed and evolving on a regular basis. The major limitation to this area of activity is the development of robust and reliable approaches to analysis. Some selected examples have been provided to assist the reader to evaluate some of the procedures that have already been developed. However, after reading this chapter you will have quickly realized that the methods are limited to only a few metal species.

References

1. United States Environmental Protection Agency, 'Acid digestion of sediments, sludges and soils', *EPA Method 350B*, National Technical Information Service, Springfield, VA, 1996.
2. United States Environmental Protection Agency, 'Acid digestion of sediments, sludges and soils', *EPA Method 350B* (Optional Method for Sb, Ba, Pb and As), National Technical Information Service, Springfield, VA, 1996.
3. Crews, H. M., Burrell, J. A. and McWeeny, D. J., *J. Sci. Food Agric.*, **34**, 997–1004 (1983).
4. Abu-Samra, A., Morris, J. S. and Koityohann, S. R., *Anal. Chem.*, **47**, 1475–1477 (1975).
5. Zlotorzynski, A., *Critical Rev. Anal. Chem.*, **25**, 43–76 (1995).
6. Jacob, J. and Boey, F., *J. Mater. Sci.*, **30**, 5321–5327 (1995).
7. Kingston, H. M. and Jassie, L. B. (Eds), *Introduction to Microwave Sample Preparation*, ACS Professional Reference Books, American Chemical Society, Washington, DC, 1988.
8. Caddick, S., *Tetrahedron*, **51**, 10403–10432 (1995).
9. United States Environmental Protection Agency, 'Microwave assisted acid digestion of sediments, sludges, soils and oils', *EPA Method 3051*, National Technical Information Service, Springfield, VA, 1994.
10. United States Environmental Protection Agency, 'Microwave assisted acid digestion of siliceous and organically based matrices', *EPA Method 3052*, National Technical Information Service, Springfield, VA, 1996.
11. Westoo, G., *Acta Chem. Scand.*, **20**, 2131–2137 (1966).
12. Quevauviller, Ph., *Method Performance Studies in Speciation Analysis*, The Royal Society of Chemistry, Cambridge, UK, 1998, pp. 44–50.
13. Quevauviller, Ph., Filippelli, M. and Horvat, M., *Trends Anal. Chem.*, **19**, 157–166 (2000).
14. Hoch, M., *Appl. Geochem.*, **16**, 719–743 (2001).
15. HMSO, *The Determination of Organic, Inorganic, Total and Specific Tin Compounds in Water, Sediments and Biota*, Her Majesty's Stationary Office, London, 1992.

16. Quevauviller, Ph., *Method Performance Studies in Speciation Analysis*, The Royal Society of Chemistry, Cambridge, UK, 1998, pp. 71–75.
17. Ebdon, L., Hill, S. J. and Rivas, C., *Trends Anal. Chem.*, **17**, 277–288 (1998).
18. Branch, S., Ebdon, L. and O'Neill, P., *J. Anal. At. Spectrom.*, **9**, 33–37 (1994).
19. Beauchemin, D., Bednas, M. E., Berman, S. S., McLaren, J. W., Siu, K. W. M. and Sturgeon, R. E., *Anal. Chem.*, **60**, 2209–2212 (1988).
20. Crews, H. M., Burrell, J. A. and McWeeny, D. J., *Z. Lebensm. Unters Forsch.*, **180**, 221–226 (1985).
21. Guerin, T., Astruc, A. and Astruc, M., *Talanta*, **50**, 1–24 (1999).
22. Quevauviller, Ph., *Method Performance Studies in Speciation Analysis*, The Royal Society of Chemistry, Cambridge, UK, 1998, pp. 131–133.
23. United States Environmental Protection Agency, 'Alkaline digestion for hexavalent chromium', *EPA Method 3060A*, National Technical Information Service, Springfield, VA, 1996.
24. United States Environmental Protection Agency, 'Chromium, hexavalent (colorimetric)', *EPA Method 7196A*, National Technical Information Service, Springfield, VA, 1992.
25. Camara, C., Cornelius, R. and Quevauviller, Ph., *Trends Anal. Chem.*, **19**, 189–194 (2000).
26. Ure, A. M., *Sci. Total Environ.*, **178**, 3–10 (1996).
27. Ure, A. M., Quevauviller, Ph., Muntau, H. and Griepink, B., *Int. J. Environ. Anal. Chem.*, **51**, 135–151 (1993).
28. Quevauviller, Ph., *Trends Anal. Chem.*, **17**, 632–642 (1998).
29. Ruby, M. V., Davies, A., Link, T. E., Schoof, R., Chaney, R. L., Freeman, G. B. and Bergstrom, P., *Environ. Sci. Technol.*, **27**, 2870–2877 (1993).
30. Ruby, M. V., Davies, A., Schoof, R., Eberle, S. and Sellstone, C. M., *Environ. Sci. Technol.*, **30**, 422–430 (1996).
31. Smith, B., Rawlins, B. G., Cordeiro, M. J. A. R., Hutchins, M. G., Tiberindwa, J. V., Sserunjogi, L. and Tomkins, A. M., *J. Geol. Sci.*, **157**, 885–891 (2000).
32. Veerabhand, M., 'Evaluation of soil extraction methods', *MSc Dissertation*, Northumbria University, Newcastle, UK, 2001.

Chapter 6

Liquids – Natural and Waste Waters

Learning Objectives

- To understand the need for separation and/or pre-concentration for metal ions in solution.
- To understand the theory of liquid–liquid extraction (LLE).
- To be able to carry out LLE in a safe and controlled manner.
- To understand the role of ion-exchange in terms of metal separation and/or pre-concentration.
- To appreciate the role of co-precipitation for pre-concentration of metal ions in solution.

6.1 Introduction

The ideal scenario in the analysis of liquid samples for trace, minor and major metals is that the analytical technique chosen to perform the analysis requires no sample pre-treatment or perhaps just a simple filtration (0.2 μm) of the sample to remove particulates prior to introduction into the chosen instrument. If this was the case, then no further need for this chapter would exist. However, reality is often very different and even the most sensitive of analytical techniques e.g. inductively coupled plasma–mass spectrometry (ICP–MS) (see Chapter 11) may require some additional sample pre-treatment, e.g. separation and/or pre-concentration. It should be remembered that any form of separation and/or pre-concentration can also have the same effect on a potential contaminant as well as on the metal of interest. Great care is required, therefore, in the choice of reagents, containers, e.g. beakers, and instrumentation used for these procedures. The most important pre-concentration methods for metal ions are liquid–liquid extraction and ion-exchange.

6.2 Liquid–Liquid Extraction

The theory of liquid–liquid extraction (LLE) is described in Box 6.1. The basis of this approach is to take a large volume of the sample, i.e. water, containing the metal of interest, and to form a metal complex, by the addition of a suitable chelating agent, which will then effectively partition into an immiscible organic solvent. The basis being that the metal present in a large volume of sample water is effectively and quantitatively transferred into a small volume of organic solvent. The most important variables in LLE are as follows:

Box 6.1 Theory of Liquid–Liquid Extraction

Two terms are used to describe the distribution of an analyte between two immiscible solvents, i.e. the *distribution coefficient* and the *distribution ratio*. The distribution coefficient is an equilibrium constant which describes the distribution of an analyte, A, between two immiscible solvents, e.g. an aqueous and an organic phase. For example, an equilibrium can be obtained by shaking the aqueous phase containing the analyte, A, with an organic phase, such as hexane. This process can be written as an equation, as follows:

$$A(aq) \rightleftharpoons A(org) \qquad (6.1)$$

where (aq) and (org) represent the aqueous and organic phases, respectively. The ratio of the activities of A in the two solvents is constant and can be represented by the following:

$$K_d = [A]_{org}/[A]_{aq} \qquad (6.2)$$

where K_d is the distribution coefficient. While the numerical value of K_d provides a useful constant, at a particular temperature, the activity coefficients are neither known or easily measured [1]. A more useful expression is the fraction of analyte extracted, E, often expressed as a percentage [2], as follows:

$$E = C_o V_o/(C_o V_o + C_{aq} V_{aq}) \qquad (6.3)$$

Continued on page 101

■ *Continued from page 100* ■

or:

$$E = K_d V/(1 + K_d V) \qquad (6.4)$$

where C_o and C_{aq} are the concentrations of the analyte in the organic phase and aqueous phases, respectively, V_o and V_{aq} are, respectively, the volumes of the organic and aqueous phases, and V is the phase ratio, i.e. V_o/V_{aq}.

For one-step liquid–liquid extractions, K_d must be large, i.e. > 10, for quantitative recovery ($>99\%$) of the analyte in one of the phases, e.g. the organic solvent [2]. This is a consequence of the phase ratio, V, which must be maintained within a practical range of values, e.g. $0.1 < V < 10$ (equation (6.4)). Typically, two or three repeat extractions are required with fresh organic solvent to achieve quantitative recoveries. The following equation (6.5) is used to determine the amount of analyte extracted after successive multiple extractions:

$$E = 1 - [1/(1 + K_d V)]^n \qquad (6.5)$$

where n is the number of extractions. For example, if the volume of the two phases are equal ($V = 1$) and $K_d = 3$ for an analyte, then four extractions ($n = 4$) would be required to achieve $>99\%$ recovery.

There can be situations where the actual chemical forms of the analyte in the aqueous and organic phases are not known, e.g. a variation in pH would have a significant effect on a weak acid or base. In this case, the distribution ratio, D, is used, as follows:

$$D = \frac{\text{concentration of A in all chemical forms in the organic phase}}{\text{concentration of A in all chemical forms in the aqueous phase}}$$

$$\qquad (6.6)$$

Note that for simple systems, when no chemical dissociation occurs, the distribution ratio is identical to the distribution coefficient.

- choice of chelating agent
- choice of organic solvent
- pH of the aqueous sample

The most commonly used chelating agent in atomic spectroscopy is ammonium pyrrolidine dithiocarbamate (APDC). This reagent, which is normally used as a 1–2% aqueous solution, can be directly extracted with methylisobutyl ketone (MIBK) as the organic solvent. As APDC forms extractable metal complexes over a wide range of pH (Table 6.1), careful control of pH allows some degree

Table 6.1 Dependence of pH on the ability of ammonium pyrrolidine dithiocarbamate (APDC) to form complexes with various metals [3]

pH range	Metals forming APDC complexes
2	W
2–4	Nb, U
2–6	As, Cr, Mo, V, Te
2–8	Sn
2–9	Sb, Se
2–14	Ag, Au, Bi, Cd, Co, Cu, Fe, Hg, Ir, Mn, Ni, Os, Pb, Pd, Pt, Ru, Rh, Tl, Zn

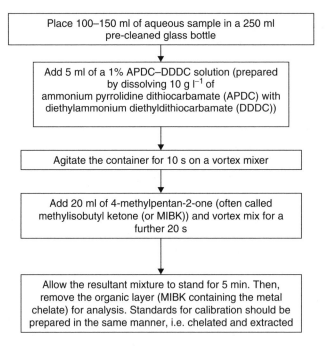

Figure 6.1 Typical procedure used for the liquid–liquid extraction of metals.

of selectivity to be introduced with this reagent. The procedure for the extraction of metals from solution using APDC into MIBK is shown in Figure 6.1.

SAQ 6.1

Is it possible to separate tin (Sn) from lead (Pb) by forming an APDC complex?

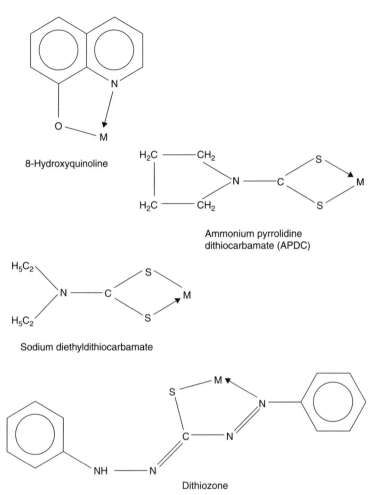

Figure 6.2 Some common metal chelates used for solvent extraction. Note that in all cases, the hydrogen ion of the parent chelating agent has been replaced by a metal.

Some other common chelating agents used in solvent extraction are shown in Figure 6.2. Of these, 8-hydroxyquinoline is particularly useful for extracting Al, Mg, Sr, V and W, diethyldithiocarbamates (e.g. the sodium derivative) for As(III), Bi, Sb(III), Se(IV), Sn(IV), Te(IV), Tl(III) and V(V), and dithiozone for Ag, Bi, Cu, Hg, Pb, Pd, Pt and Zn.

6.3 Ion-Exchange

In this case, cation- or anion-exchange chromatographic techniques are commonly used. As their names suggest, cation-exchange is used to separate metal ions (positively charged species), while anion-exchange is employed to separate negatively charged species. At first hand, it may seem that the only useful form of ion-exchange chromatography for metal separation/determination is cation-exchange, but this is not always the case.

DQ 6.1

Can you suggest any cases where anion-exchange chromatography may be used in metal determination?

Answer

Some elements are found as their anions. For example, SO_4^{2-} can be used for sulfur determination, while both arsenite (AsO_2^{-}) and arsenate (AsO_4^{3-}) can be used for arsenic determination.

As an example, in the use of a strong cation-exchange resin, the following general equations can be written:

1. Metal ion (M^{n+}) pre-concentrated on cation exchange resin

$$n\text{RSO}_3^-\text{H}^+ + M^{n+} = (\text{RSO}_3^-)nM^{n+} + \text{H}^+$$

2. Desorption of metal ion using acid

$$(\text{RSO}_3^-)nM^{n+} + \text{H}^+ = n\text{RSO}_3^-\text{H}^+ + M^{n+}$$

An alternative approach which allows the separation of an excess of alkali metal ions from other cations uses a chelating ion-exchange resin. This type of resin forms chelates with the metal ions. The most common of these is 'Chelex-100'. This resin contains iminodiacetic acid functional groups which behave in a similar way to ethylenediaminetetraacetic acid (EDTA). It has been found that 'Chelex-100', in acetate buffer at pH 5–6, can retain Al, Bi, Cd, Co, Cu, Fe, Ni, Pb, Mn, Mo, Sc, Sn, Th, U, V, W, Zn and Y, plus various rare-earth metals, while at the same time it does not retain alkali metals (e.g. Li, Na, Rb and Cs), alkali-earth metals (Be, Ca, Mg, Sr and Ba) and anions (F^-, Cl^-, Br^- and I^-).

By using ion-exchange resins, a selected metal ion can be isolated (separated) and pre-concentrated from its matrix. This process can be carried out in two ways, i.e. by (i) batch, or (ii) column processes. In the batch process, the ion-exchange resin is added to the aqueous sample, whereas in the second process the resin is packed into a chromatographic column. The former would be always carried out off-line, while the latter could be carried out in either on-line or off-line mode.

6.4 Co-Precipitation

While other techniques, such as co-precipitation, can be used for pre-concentration they are not as common as those discussed above. Co-precipitation allows the quantitative precipitation of the metal ion of interest by the addition of a co-precipitant.

Co-precipitation of metal ions on collectors can be attributed to several mechanisms, as follows:

- *adsorption* – the charge on the surface can attract ions in solution of opposite charge
- *occlusion* – ions are embedded within the forming precipitate
- *cocrystallization* – the metal ion can become incorporated in the crystal structure of the precipitate

The major disadvantage of co-precipitation is that the precipitate, which is present at a high mass-to-analyte ratio, can be a major source of contamination. In addition, further sample preparation, e.g. dissolution, is required prior to analysis for the analyte. This can also increase the risk of contamination and analyte losses. One of the most common co-precipitants is iron, e.g. as $Fe(OH)_3$.

6.5 Summary

The presence of trace metals in natural and waste waters can often cause a problem in terms of the selected analytical technique. In order to be able to quantify the concentration of trace metals in aqueous samples, appropriate methods of pre-concentration therefore need to be selected. This present chapter has summarized the main methods available for such pre-concentration procedures.

References

1. Cresser, M. S., *Solvent Extraction in Flame Spectroscopic Analysis*, Butterworths, London, 1978.
2. Majors, R. E., *LC–GC Int.*, **10**, 93–101 (1997).
3. Kirkbright, G. F. and Sargent, M., *Atomic Absorption and Fluorescence Spectroscopy*, Academic Press, London, 1974.

Sample Preparation for Organic Analysis

Chapter 7

Solids

Learning Objectives

- To appreciate the different approaches available for the preparation of solid samples for organic analysis.
- To be able to carry out Soxhlet extraction in a safe and controlled manner.
- To be able to carry out shake-flask extraction in a safe and controlled manner.
- To be able to carry out ultrasonic extraction in a safe and controlled manner.
- To understand the important properties of a supercritical fluid.
- To appreciate the instrumental requirements for supercritical fluid extraction (SFE).
- To be able to carry out SFE in a safe and controlled manner.
- To appreciate the instrumental requirements for microwave-assisted extraction (MAE).
- To be able to carry out MAE in a safe and controlled manner.
- To understand the theory relating to pressurized fluid extraction (PFE).
- To appreciate the instrumental requirements for PFE.
- To be able to carry out PFE in a safe and controlled manner.
- To appreciate the requirements for matrix solid-phase dispersion (MSPD).
- To be able to carry out MSPD in a safe and controlled manner.

7.1 Introduction

Extraction of organic pollutants from solid or semi-solid matrices has been widely investigated in recent years. New instrumental approaches have led to developments in terms of speed of extraction, reduction in organic solvent consumption and the introduction of automation. However, these instrumental approaches

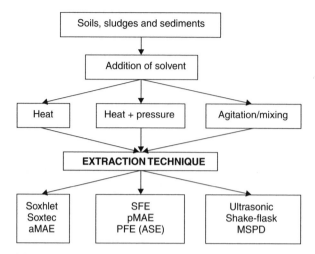

Figure 7.1 Various methods of extraction of analytes from solid matrices.

come with a higher price tag than the conventional approaches. The conventional approaches are normally based on larger volumes of organic solvent, glassware and simpler forms of heating. They can also be described as very labour-intensive and relatively slow. Nevertheless, they have been available for many years and are often readily available in the average laboratory.

Essentially, extraction of environmental pollutants from solid or semi-solid matrices can be divided into several categories based on the method of extraction, mode of heating and presence or not of some type of agitation. This is demonstrated in Figure 7.1.

SAQ 7.1

From an examination of Figure 7.1, which of the techniques shown would you identify as *conventional* techniques and which as *instrumental* approaches?

7.2 Soxhlet Extraction

Soxhlet extraction is used as the benchmark against which any new extraction technique is compared. The basic Soxhlet extraction apparatus consists of a solvent reservoir, an extraction body, a heat source (e.g. an isomantle) and a water-cooled reflux condenser (Figure 7.2). A Soxhlet uses a range of organic solvents to remove organic compounds, primarily from solid matrices. The solid sample (ca. 10 g if a soil) and a similar mass of anhydrous sodium sulfate are placed in the porous thimble (cellulose), which in turn is located in the inner tube of the Soxhlet apparatus. The apparatus is then fitted to a round-bottomed flask

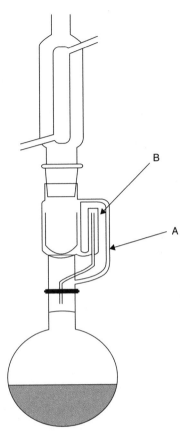

Figure 7.2 The basic Soxhlet extraction system. From Dean, J. R., *Extraction Methods for Environmental Analysis*, Copyright 1998. © John Wiley & Sons Limited. Reproduced with permission.

of appropriate volume containing the organic solvent of choice, and to a reflux condenser. The solvent is then boiled gently using an isomantle – the solvent vapour passes up through the tube marked (A), is condensed by the reflux condenser, and the condensed solvent falls into the thimble and slowly fills the body of the Soxhlet apparatus. When the solvent reaches the top of the tube (B), it syphons over into the round-bottomed flask the organic solvent containing the analyte extracted from the sample in the thimble. The solvent is then said to have completed one cycle.

DQ 7.1

Based on a solvent volume of 150 ml, how long would you estimate one cycle takes?

Answer

One cycle takes approximately 15 min, i.e. about four cycles per hour.

The whole process is repeated frequently until the pre-set extraction time is reached. As the extracted analyte will normally have a higher boiling point than the solvent, it is preferentially retained in the flask and fresh solvent recirculates. This ensures that only fresh solvent is used to extract the analyte from the sample in the thimble. A disadvantage of this approach is that the organic solvent is below its boiling point when it passes through the sample contained in the thimble. In practice, this is not necessarily a problem as Soxhlet extraction is normally carried out over long time-periods, i.e. 6, 12, 18 or 24 h.

SAQ 7.2

In terms of percentage efficiency at removing an analyte from a matrix, what would you envisage the 'extraction efficiency–time profile' to look like?

A typical procedure for Soxhlet extraction is shown in Figure 7.3.

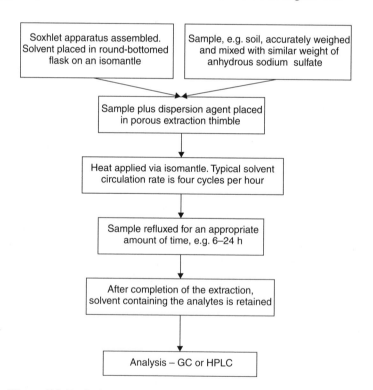

Figure 7.3 Typical procedure used for the Soxhlet extraction of solids.

Automated Soxhlet extraction, or *Soxtec*, utilizes a three-stage process to obtain rapid extractions. In Stage 1, a thimble containing the sample is immersed in the boiling solvent for approximately 60 min. After this, the thimble is elevated above the boiling solvent and the sample extracted as in the standard Soxhlet extraction approach (Stage 2). This is carried out for up to 60 min. The final stage (Stage 3) involves the evaporation of the solvent directly in the Soxtec apparatus (10–15 min). This approach has several advantages, including speed (it is faster than normal Soxhlet extraction, i.e. approximately 2 h per sample), the fact that it uses only 20% of the solvent volume of Soxhlet extraction, and the sample can be concentrated directly in the Soxtec apparatus.

7.2.1 Example 7.1: Soxhlet Extraction of Polycyclic Aromatic Hydrocarbons from Contaminated Soil

7.2.1.1 Extraction Conditions

These were as follows:

- Sample: 10 g, plus 10 g anhydrous sodium sulfate
- Solvent: 150 ml dichloromethane
- Extraction time: 24 h

Comments The sample was heated by using an isomantle. Typically, refluxing of the solvent occurred at the rate of four cycles per hour. Extracts were concentrated to 10 ml using a rotary evaporator and then diluted twofold before addition of the internal standards.

7.2.1.2 Analysis of Extracts by GC–MS

Separation and identification of the individual polycyclic (polynuclear) aromatic hydrocarbons (PAHs) was carried out on an HP 5890 Series II Plus gas chromatograph, fitted to an HP 5972A mass spectrometer. A 30 m $\times$ 0.25 mm id $\times$ 0.25 μm film thickness DB-5 capillary column was used, with temperature programming from an initial temperature of 85°C held for 2 min, before commencing a 6°C min^{-1} rise to 300°C, to give a final time of 7 min. The split/splitless injector was held at 300°C and operated in the splitless mode with the split valve closed for 1 min following sample injection. The split flow was set at 40 ml min^{-1}, and the mass spectrometer transfer line was maintained at 270°C. Electron impact ionization at 70 eV, with the electron multiplier voltage set at 1500 V, was used, while operating in the single-ion monitoring (SIM) mode.

7.2.1.3 Typical Results

These are shown in Figure 7.4 [1].

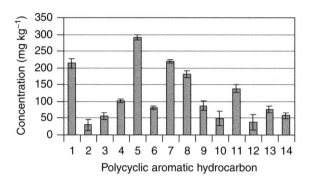

Figure 7.4 Results obtained for the Soxhlet extraction of various polycyclic aromatic hydrocarbons from contaminated soil: 1, naphthalene; 2, acenaphthylene; 3, acenaphthene; 4, fluorene; 5, phenanthene; 6, anthracene; 7, fluoranthene; 8, pyrene; 9, benz[*a*]anthracene; 10, chrysene; 11, benzo[*b,k*]fluoranthene; 12, benzo[*a*]pyrene; 13, indeno[1,2,3-*cd*]pyrene; 14, benzo[*ghi*]pyrene [1] (cf. DQ 7.2).

DQ 7.2

Comment on the results obtained in this study (see Figure 7.4).

Answer

It can be seen that Soxhlet extraction is able to extract a whole range of polycyclic aromatic hydrocarbons effectively and with reasonable precision.

7.3 Shake-Flask Extraction

Conventional liquid–solid extraction, in the form of shake-flask extraction, is carried out by placing a soil sample into a suitable glass container, adding a suitable organic solvent, and then agitating or shaking.

DQ 7.3

What agitating or shaking actions do you think are possible?

Answer

The most common actions are rocking, a circular action or an 'end-over-end' motion.

Agitating or shaking is carried out for a pre-specified time-period. After extraction, the solvent containing the analyte needs to be separated from the matrix by means of centrifugation and/or filtration. In some instances, it may be advisable to repeat the process several times with fresh solvent and then combine all

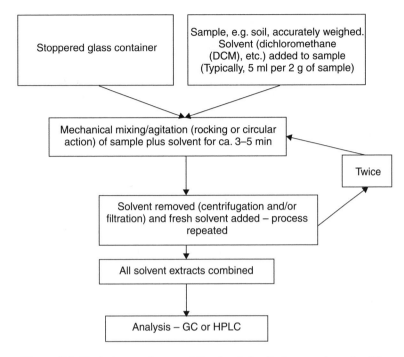

Figure 7.5 Typical procedure used for the shake-flask extraction of solids.

of the extracts. A typical procedure used for shake-flask extraction is shown in Figure 7.5.

7.3.1 Example 7.2: Shake-Flask Extraction of Phenols from Contaminated Soil

7.3.1.1 Extraction Conditions

These were as follows:

- Sample: 1 g
- Solvent: 50 ml methanol–water (60–40 vol%)
- Extraction time: 30 min

Comments The sample and solvent were placed in a 100 ml screw-capped bottle and extracted on a rotating-disc Warburg mixer. The resultant sample/solvent mixture was filtered under vacuum, and the extracted sample then filtered through a 0.45 μm membrane 'Acrodisk' prior to analysis.

7.3.1.2 Analysis by HPLC

Separation and quantitation was achieved by using a 25 cm × 4.6 mm id ODS2 column with UV detection at 275 nm. The mobile phase was acetonitrile–H_2O–

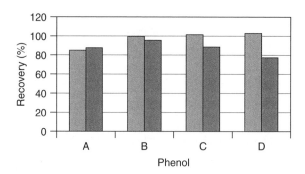

Figure 7.6 Results obtained for the shake-flask extraction of various phenols from contaminated soil and 'Celite': A, phenol; B, 3-cresol; C, 4-ethylphenol; D, 1-naphthol: ■, 'Celite'; ■, soil [2] (cf. DQ 7.4).

acetic acid (40 : 591), operating under isocratic conditions, at a flow rate of 1 ml min^{-1}. A 20 μl 'Rheodyne' injection loop was used to introduce samples and standards onto the column (at 30°C).

7.3.1.3 Typical Results

These are shown in Figure 7.6 [2].

DQ 7.4

Comment on the results obtained in this study (see Figure 7.6).

Answer

It can be noted that, in most cases, lower recoveries are obtained when shake-flask extraction is used to remove phenols from soil when compared to 'Celite' (an inert siliceous matrix). This effect is most pronounced with 1-naphthol.

7.4 Ultrasonic Extraction

Sonication involves the use of sound waves to agitate a sample immersed in an organic solvent. The preferred approach is to use a sonic probe, although a sonic bath can also be used. The sample is placed in a suitable glass container and enough organic solvent is then added to cover the sample. The system is then sonicated for a short time-period, typically 3 min, using the sonic bath or probe. After extraction, the solvent containing the analyte is separated by centrifugation and/or filtration and fresh solvent added. The whole process is repeated three times and all of the solvent extracts are then combined.

DQ 7.5

Why is it better to extract with fresh solvent each time?

Answer

Extraction is a partitioning process between the analyte, matrix and solvent. In order to allow an analyte to partition successfully, new solvent is required. This allows the target analyte to effectively partition in the fresh solvent. Repeating the entire process three times therefore allows maximum recovery of analyte, compared with two extractions. It is likely that four extractions will produce even more analyte. However, the extra time and effort required to perform a fourth extraction is outweighed by the small recovery of analyte expected.

A typical procedure used for ultrasonic extraction is shown in Figure 7.7.

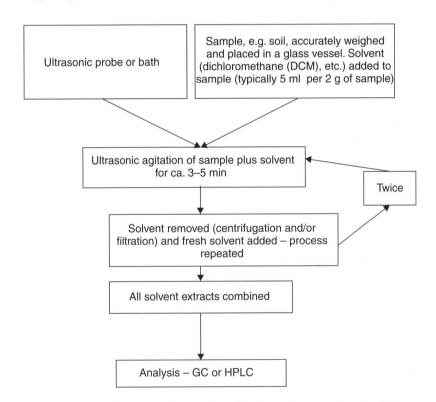

Figure 7.7 Typical procedure used for the ultrasonic extraction of solids.

7.5 Supercritical Fluid Extraction

Supercritical fluid extraction (SFE) relies on the diversity of properties exhibited by a supercritical fluid to (selectively) extract analytes from solid, semi-solid or liquid matrices. The important properties offered by a supercritical fluid for extraction are (i) good solvating power, (ii) high diffusivity and low viscosity, and (iii) minimal surface tension. For background information on supercritical fluids, see Box 7.1.

Box 7.1 Supercritical Fluids

History

The term *supercritical fluid* is used to describe any substance above its critical temperature and pressure. The discovery of the supercritical phase is attributed to Baron Cagniard de la Tour in 1822 [3]. He observed that the boundary between a gas and a liquid disappeared for certain substances when the temperature was increased in a sealed glass container. While some further work was carried out on supercritical fluids, the subject remained essentially dormant until 1964 when a patent was filed for using supercritical carbon dioxide to decaffeinate coffee. Subsequent major developments by food manufacturers have led to the commercialization of this approach in coffee production. The use of supercritical fluids in the laboratory was initially focused on their use in chromatography, particularly capillary supercritical fluid chromatography (SFC). However, it was not until the mid-1980s that the use of SFE for extraction was commercialized.

Definition of a Supercritical Fluid

A phase diagram for a pure substance is shown in Figure 7.8. In this diagram can be seen the regions where the substance occurs, as a consequence of temperature or pressure, as a single phase, i.e. solid, liquid or gas. The divisions between these regions are bounded by curves indicating the co-existence of two phases, e.g. solid–gas, corresponding to sublimation, solid–liquid, corresponding to melting, and finally, liquid–gas, corresponding to vaporization. The three curves intersect at the triple point where the three phases co-exist in equilibrium. At the critical point, designated by both a critical temperature and a critical pressure, no liquefaction will take place on raising the pressure and no gas will be formed on increasing the temperature. It is this defined regime, which is by definition, the *supercritical region*. The critical properties for some selected substances are shown in Table 7.1.

Continued on page 119

Continued from page 118

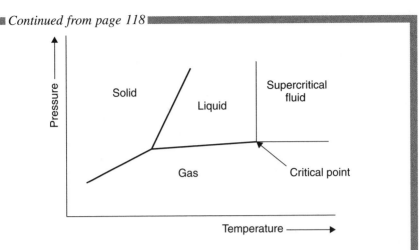

Figure 7.8 Phase diagram for a pure substance. From Dean, J. R., *Extraction Methods for Environmental Analysis*, Copyright 1998. © John Wiley & Sons Limited. Reproduced with permission.

Table 7.1 Critical properties of some selected materials. From Dean, J. R., *Extraction Methods for Environmental Analysis*, Copyright 1998. © John Wiley & Sons Limited. Reproduced with permission

Substances	Critical temperature (°C)	Critical pressure (atm)	Critical pressure (psi)
Ammonia	132.4	115.0	1646.2
Carbon dioxide	31.1	74.8	1070.4
Chlorodifluoromethane	96.3	50.3	720.8
Ethane	32.4	49.5	707.8
Methanol	240.1	82.0	1173.4
Nitrous oxide	36.6	73.4	1050.1
Water	374.4	224.1	3208.2
Xenon	16.7	59.2	847.0

DQ 7.6

Why do you think that carbon dioxide is the supercritical fluid of choice?

Answer

The low critical temperature (31°C) and moderate critical pressure (74.8 atm) of carbon dioxide have made it the prominent substance. It is also relatively safe to use and non-toxic (unless of course you filled the entire room with this gas!).

7.5.1 *Instrumentation*

An SFE system consists of six basic components, namely a supply of high-purity carbon dioxide (CO_2), a supply of high-purity organic modifier, two pumps, the oven for the extraction cell, the pressure outlet or restrictor, and the collection vessel. A schematic diagram of a typical SFE arrangement is shown in Figure 7.9. The CO_2 is supplied in a cylinder fitted with a dip tube which allows only lique-fied CO_2 to be accessible. This is in contrast to the common cylinder type which comprises liquid CO_2 at the bottom and vapour at the top (when positioned verti-cally). The great disadvantage of CO_2 is its lack of a permanent dipole moment. This limits the applicability of SFE to the extraction of non- or moderately polar analytes. Therefore, an organic solvent, a so-called modifier, needs to be added in order to generate a more polar solvent mixture. Modified CO_2 can be produced via the use of two pumps or via a single cylinder containing CO_2 and an organic modifier. The choice of approach also has a financial consideration. The use of two pumps increases the capital cost of the instrumentation, while the use of mixed cylinders raises the operating costs of the system. Scientifically, the use of two pumps is to be preferred.

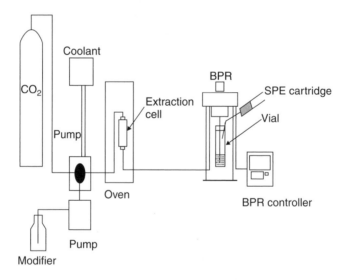

Figure 7.9 Schematic of the layout of a typical supercritical fluid extraction system: BPR, back-pressure regulator (restrictor); SPE, solid-phase extraction. From Dean, J. R., *Extraction Methods for Environmental Analysis*, Copyright 1998. © John Wiley & Sons Limited. Reproduced with permission.

DQ 7.7

Why do you think that the use of a single, i.e. 'mixed' cylinder, is not the preferred option?

Answer

The 'mixed cylinder' has its shortcoming in that the CO_2-to-modifier ratio varies over the lifetime of its operation. It therefore does not deliver 'what it says on the label'!

Two pumps are required to pump the pressurized CO_2 and organic modifier through the SFE system. After pumping out from the cylinder, CO_2 and organic modifier are mixed with a 'T-piece' before introducing into the extraction cell. The combination of the two pumps allows a degree of control in terms of modifier composition (1–20 vol%) and flexibility of solvent choice. The ideal pump should deliver a constant flow rate (in the ml min^{-1} range) at a suitable pressure (3500–10 000 psi). Two kinds of pumps, the reciprocating type (piston pump) and syringe pump, are usually used in SFE systems, although the former is the most commonly used due to its lower cost. However, when the reciprocating pump is employed, cooling is required to prevent *cavitation*, i.e. gas entrapment, in the pump head. No such modification of the pump head is obviously required for the organic solvent (modifier).

The external heating, to the solvent system, needed to generate the critical temperature is carried out via an oven in which is located the sample cell or vessel. Generally, the required temperature range of the oven is up to 100°C. However, temperatures of 200–250°C may be advantageous in some circumstances. The sample vessel, which is typically made of stainless-steel, must be capable of safely withstanding high pressures (up to 10 000 psi). It should be remembered that some time will be needed to achieve the pre-set temperature for the newly prepared and inserted sample-containing vessel prior to starting the extraction process. Almost all commercial extraction vessels are of the flow-through design, which allows fresh, clean supercritical fluid to pass over the sample. Many commercial extraction vessels are capable of insertion into the system, without the need for wrenches, to allow ease of use and rapid changeover of samples. This is very helpful in preventing the pressure fittings from suffering excessive wear and tear.

Either fixed or variable (mechanical or electronically controlled) restrictors have been employed to maintain the pressure within the extraction vessel. The former is typified by the use of narrow fused-silica or metal capillary tubing, with the latter by back-pressure regulators (BPRs). However, the fixed restrictor is not a good choice, on account of its lack of robustness, although it is the cheapest

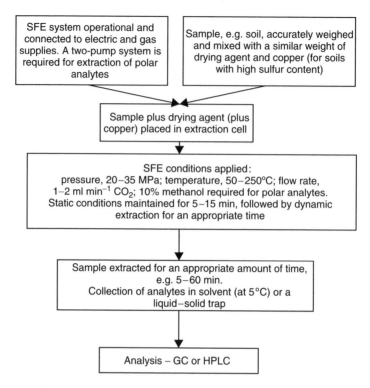

Figure 7.10 Typical procedure used for the supercritical fluid extraction of solids.

option. For the extraction of real samples, it is more appropriate to use a variable restrictor. A typical procedure used for supercritical fluid extraction is shown in Figure 7.10.

7.5.2 Example 7.3: Supercritical Fluid Extraction of Organochlorine Pesticides from Contaminated Soil and 'Celite'

7.5.2.1 Extraction Conditions

These were as follows:

- Sample: 1 g

- SFE conditions: pressure, 250 kg cm^{-2}; temperature, 50°C; static extraction time, 15 min, followed by 40 min dynamic extraction time; flow rate of liquid CO_2, 2 ml min^{-1}

Comments Extracts collected in a vial containing 3–4 ml dichloromethane (DCM). Escaping CO_2 and analytes vented through a C18 SPE cartridge which was back-flushed with 1–2 ml methanol after each extraction.

7.5.2.2 Analysis by GC–MS

Separation and identification of the individual organochlorine pesticides (OCPs) was carried out on an HP 5890 Series II Plus gas chromatograph, fitted to an HP 5972A mass spectrometer. A 30 m × 0.25 mm id × 0.25 μm film thickness DB-5 capillary column was used, with temperature programming from an initial temperature held at 85°C for 0.75 min before commencing a 16°C min^{-1} rise to 285°C, with a final time of 2 min. The split/splitless injector was held at 280°C and operated in the splitless mode with the split valve closed for 1 min following sample injection. The split flow was set at 40 ml min^{-1}, and the mass spectrometer transfer line was maintained at 290°C. Electron impact ionization at 70 eV, with the electron multiplier voltage set at 1500 V, was used, while operating in the single-ion monitoring (SIM) mode.

7.5.2.3 Typical Results

These are shown in Figure 7.11 [4].

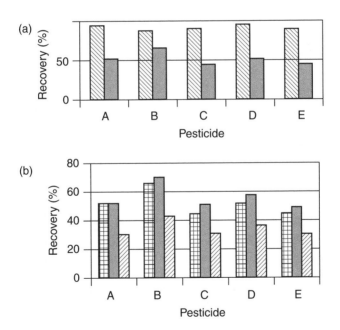

Figure 7.11 Results obtained for the supercritical extraction of various organochlorine pesticides from contaminated soil and 'Celite', showing the influence of (a) the soil matrix, and (b) the soil organic matter (SOM) content: A, lindane; B, aldrin; C, dieldrin; D, heptachlor; E, isodrin: (a) ▨, 'Celite'; ■, soil: (b) ▦, SOM 0.2%; ■, SOM 15%; ▨, SOM 35% [4] (cf. DQ 7.8).

DQ 7.8

Comment on the results obtained in this study (see Figure 7.11).

Answer

In the first situation, (see Figure 7.11(a)), it is noted that the recovery of organochlorine pesticides is influenced by the soil itself. The highest recoveries are obtained from 'Celite', an inert siliceous matrix. This indicates that SFE suffers from some matrix dependency. This behaviour is further investigated in the second situation (Figure 7.11(b)). In this case, the matrix dependency is found to be soil organic matter (SOM) related. It is found that the highest recoveries of organochlorine pesticides are obtained when the soil organic matter is low (15% or less) and vice versa.

7.6 Microwave-Assisted Extraction

Microwave-assisted extraction (MAE) utilizes organic solvent and heat to extract organic pollutants from solid matrices. The major difference between this approach and others is the use of a microwave oven as the heat source. For background information on microwave ovens, see Box 5.1.

7.6.1 Instrumentation

At present, a number of manufacturers supply microwave ovens which are specially designed for laboratory use. However, during the initial period to develop microwave-assisted extraction, domestic microwave ovens were often employed in laboratories for this purpose. The dedicated instruments possess several advantages over domestic ovens, particularly in terms of safety features, although they are considerably more expensive. Generally, the dedicated microwave ovens can be divided into either a closed-vessel style or an open-vessel style. The former is typified by the 'MARS 5' system, supplied by the CEM Corporation, USA, and the latter by the 'Soxwave' system, from Prolabo Ltd, France (now the CEM Corporation). In addition, some laboratory microwave ovens, such as the QLAB 6000 model, developed by the Questron Technologies Corporation, Canada, can be set for closed- or open-vessel operation, or for the flow-through mode. Schematic diagrams of atmospheric (aMAE) and pressurized (pMAE) systems are shown in Figures 5.5 and 5.6, respectively.

The MARS 5 system allows up to 14 extraction vessels (XP-1500 Plus™) to be irradiated simultaneously. In addition, other features include a function for monitoring both pressure and temperature, and most notably, the system is equipped with a solvent alarm to call attention to an unexpected release of flammable and toxic organic solvent. The microwave energy output of this system is 1500 W at a frequency of 2450 MHz (at 100% power). Pressure (up to 800 psi) is continuously measured (measurements being taken at the rate of 200 s^{-1}), while the temperature (up to 300°C) is monitored for all vessels every 7 s. All of the sample vessels are held in a carousel which is located within the microwave cavity. Each vessel has a vessel body and an inner liner. The liner is made of 'TFM' fluoropolymer and has a volume of 100 ml. A patented safety system (AutoVent Plus™) allows venting of excess pressure within each extraction vessel. The system works by lifting of the vessel cap(s) to release excess pressure and then immediately resealing to prevent loss of sample. If solvent leaking from the extraction vessel(s) does occur, a solvent monitoring system will automatically shut-off the magnetron, while still allowing the exhaust fan to continue working.

The Soxwave system possesses a microwave energy output of 300 W at full power. It can operate at percentage power increments from 0–100% and for different time intervals. As with other open-vessel systems, it deals with individual samples sequentially. Its sample vessel is a glass container, which looks like a large boiling/test tube, into which the sample is introduced, followed by the selected solvent. The sample-containing vessel is positioned within a protective glass sheath and is combined with an air or water condenser to prevent loss of volatile analytes and solvent. Commonly, the open-vessel system does not normally have a temperature-monitoring function.

Table 7.2 lists some of the characteristics of the solvents which are most commonly used in MAE.

DQ 7.9

Why is hexane on its own not a good solvent for MAE?

Answer

In order for microwave heating to occur, the solvent must have a permanent dipole moment. This can be assessed by considering a solvent's

Table 7.2 Characteristics of solvents commonly used in microwave-assisted extraction. From Dean, J. R., *Extraction Methods for Environmental Analysis*, Copyright 1998. © John Wiley & Sons Limited. Reproduced with permission

Solvent	Dielectric constant	Boiling point (°C)	Closed-vessel temperature (°C)[a]
Acetone	20.7	56.2	164
Acetonitrile	37.5	81.6	194
Dichloromethane	8.93	39.8	140
Hexane	1.89	68.7	—
Hexane–acetone	—	52.0[b]	156
Methanol	32.63	64.7	151

[a]Measured at 175 psi.
[b]Determined experimentally.

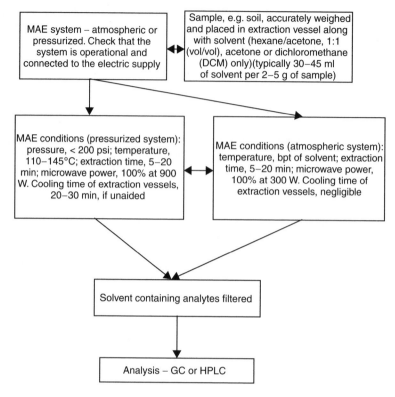

Figure 7.12 Typical procedure used for the microwave-assisted extraction of solids.

dielectric constant. It is observed (Table 7.2) that hexane has a small dielectric constant and therefore is unsuitable on its own as a solvent for MAE.

A typical procedure used for microwave-assisted extraction is shown in Figure 7.12.

7.6.2 Example 7.4: Atmospheric Microwave-Assisted Extraction of Polycyclic Aromatic Hydrocarbons from Contaminated Soil

7.6.2.1 Extraction Conditions

These are as follows:

- Sample: 2 g
- Solvent: 70 ml dichloromethane
- aMAE conditions: power, 99% (for a 300 W system); extraction time, 20 min

Comments Contents of extraction vessel filtered through a GF/A glass microbore filter. Extracts were concentrated to 5 ml using a rotary evaporator, before addition of internal standards.

7.6.2.2 Analysis by GC

Separation and identification of the individual PAHs was carried out on a Carlo Erba HRGC 5300 Mega Series gas chromatograph, with on-column injection and flame ionization detection. A 30 m × 0.32 mm id × 0.1 μm film thickness DB-5 HT capillary column was used with temperature programming from an initial temperature held at 50°C for 2 min before commencing a 15°C min^{-1} rise to 90°C; the latter temperature is held for 2 min, and then increased at a rate of 6°C min^{-1} to 300°C, with a final hold time of 8 min. The detector temperature was set at 290°C.

7.6.2.3 Typical Results

These are shown in Figure 7.13 [1].

DQ 7.10

Comment on the results obtained in this study (see Figure 7.13).

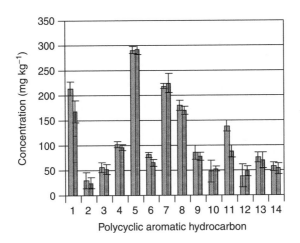

Figure 7.13 Results obtained for the atmospheric microwave-assisted extraction of various polycyclic aromatic hydrocarbons from contaminated soil, and comparison with those obtained from Soxhlet extraction: ■, Soxhlet; ▦, aMAE: 1, naphthalene; 2, acenaphthylene; 3, acenaphthene; 4, fluorene; 5, phenanthene; 6, anthracene; 7, fluoranthene; 8, pyrene; 9, benz[*a*]anthracene; 10, chrysene; 11, benzo[*b,k*]fluoranthene; 12, benzo[*a*]pyrene; 13, indeno[1,2,3-*cd*]pyrene; 14, benzo[*ghi*]pyrene [1] (cf. DQ 7.10).

Answer

It is found that the recoveries of polycyclic aromatic hydrocarbons from soil are similar irrespective of the extraction method used. In addition, similar precision is achieved in both cases.

7.6.3 Example 7.5: Pressurized Microwave-Assisted Extraction of Polycyclic Aromatic Hydrocarbons from Contaminated Soil

7.6.3.1 Extraction Conditions

These were as follows:

- Sample: 2 g
- Solvent: 40 ml acetone
- pMAE conditions: power, 30% (for a 950 W system); temperature, 120°C; extraction time, 20 min

Comments Extraction vessels allowed to cool after extraction process. Contents of vessels were then filtered through a GF/A glass microbore filter, and extracts concentrated to 5 ml by using a rotary evaporator before the addition of internal standards.

7.6.3.2 Analysis by GC

Separation and identification of the individual PAHs was carried out on a Carlo Erba HRGC 5300 Mega Series gas chromatograph, with on-column injection and flame ionization detection. A 30 m × 0.32 mm id × 0.1 μm film thickness DB-5 HT capillary-column was used with temperature programming from an initial temperature held at 50°C for 2 min before commencing a 15°C min^{-1} rise to 90°C; the latter temperature is held for 2 min, and then increased at a rate of 6°C min^{-1} to 300°C, with a final hold time of 8 min. The detector temperature was set at 290°C.

7.6.3.3 Typical Results

These are shown in Figure 7.14 [1].

DQ 7.11

Comment on the results obtained in this study (see Figure 7.14).

Answer

As already observed for atmospheric microwave-assisted extraction (see Figure 7.13 above), it is found that the recoveries of polycyclic aromatic hydrocarbons from soil are similar irrespective of the extraction method used. In addition, similar precision is achieved in both cases.

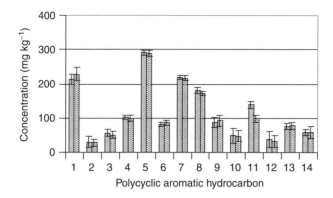

Figure 7.14 Results obtained for the pressurized microwave-assisted extraction of various polycyclic aromatic hydrocarbons from contaminated soil, and comparison with those obtained from Soxhlet extraction: ▨, Soxhlet; ▨, pMAE: 1, naphthalene; 2, acenaphthylene; 3, acenaphthene; 4, fluorene; 5, phenanthene; 6, anthracene; 7, fluoranthene; 8, pyrene; 9, benz[*a*]anthracene; 10, chrysene; 11, benzo[*b,k*]fluoranthene; 12, benzo[*a*]pyrene; 13, indeno[1,2,3-*cd*]pyrene; 14, benzo[*ghi*]pyrene [1] (cf. DQ 7.11).

7.7 Pressurized Fluid Extraction

Pressurized fluid extraction uses heat and pressure to extract analytes rapidly and efficiently from solid matrices. For background information on pressurized fluid extraction, see Box 7.2.

Box 7.2 Pressurized Fluid Extraction

Theory

Liquid solvents at elevated temperatures and pressures should provide enhanced extraction capabilities when compared to their use at or near room temperature and atmospheric pressure for two main reasons, namely (i) solubility and mass-transfer effects, and (ii) disruption of surface equilibria [5].

Solubility and Mass-Transfer Effects

The following three factors are considered important:

- Higher temperatures increase the capacity of solvents to solubilize analytes.

Continued on page 130

■ *Continued from page 129* ■

- Faster diffusion rates occur as a result of increased temperatures.
- Improved mass transfer, and hence increased extraction rates, occur when fresh solvent is introduced, i.e. the concentration gradient is greater between the fresh solvent and the surface of the sample matrix.

Disruption of Surface Equilibria

As both temperature and pressure are important, both are discussed separately.

Temperature Effects

- Increased temperatures can disrupt the strong solute–matrix interactions caused by van der Waals forces, hydrogen bonding, and dipole attractions of the solute molecules and active sites on the matrix.
- Decreases in the viscosities and surface tensions of solvents occur at higher temperatures, thus allowing an improved penetration of the matrix, and hence improved extraction.

Pressure Effects

- The utilization of elevated pressures allows solvents to remain liquified above their boiling points.
- Extraction from within the matrix is possible, as the pressure allows the solvent to penetrate the sample matrix.

7.7.1 Instrumentation

Since pressurized fluid extraction (PFE), also known as accelerated solvent extraction (ASE™), is a relatively new technique, the commercial availability of PFE instruments is limited. A commercial PFE system ('ASE 200') currently available is a fully automated sequential extractor developed by the Dionex Corporation, USA. This mainly consists of a solvent-supply system, extraction cell, oven, collection system and purge system, all of which are under computer control. A schematic diagram of a PFE system is shown in Figure 7.15. This system (ASE 200) can operate with up to 24 sample-containing extraction vessels and up to 26 collection vials, plus an additional four vial positions for rinse/waste collection.

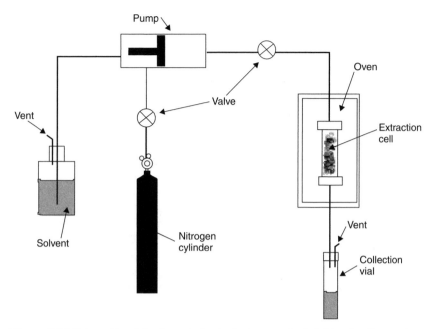

Figure 7.15 Schematic of the layout of a typical pressurized fluid extraction system.

DQ 7.12

What advantage does automation offer?

Answer

It allows samples, once loaded into the carousel, to be sequentially extracted without the interference or requirement for laboratory personnel. This allows the analyst to perform other tasks while the extraction process is being carried out.

Five sample extraction vessel sizes are available, i.e. 1, 5, 11, 22 and 33 ml. The sample vessel has removable end-caps that allow for ease of cleaning and sample filling. Each sample vessel is fitted with two finger-tight caps with compression seals for high-pressure closure. To fill a sample vessel, one end-cap is introduced and screwed on to finger-tightness. Then, a filter paper (Whatman, Grade D28, 1.98 cm diameter) is introduced into the sample vessel, followed by the sample itself. Samples should normally be air-dried or mixed with a suitable drying agent, e.g. anhydrous sodium sulfate, and ground to 100–200 mesh size (150 μm to 75 μm). Then, the other end-cap is screwed on to finger-tightness and the sample vessel placed in the carousel. Through computer control, the carousel introduces selected extraction vessels consecutively. An auto-seal actuator places the extraction vessel into the system and then returns the vessel to the carousel after extraction. The system is operated as follows. The sample cell, positioned vertically, is filled from top to bottom with the selected solvent or solvent mixture from a pump (capable of operating at up to 70 ml min^{-1}) and then heated to a designated temperature (40–200°C) and pressure (6.9–20.7 MPa or 1000–3000 psi). These operating conditions are maintained for a pre-specified time using static valves. After the appropriate time (typically 5 min), the static valves are released and a few ml of clean solvent (or solvent mixture) is passed through the sample cell to exclude the existing solvent(s) and extracted analytes. The flush volume is normally 0.6 times the cell volume. This rinsing is enhanced by the passage of N_2 gas (45 s at 150 psi) to purge both the sample cell and the stainless-steel transfer lines. After gas purging, all extracted analytes and solvent(s) are passed through 30 cm of stainless-steel tubing which punctures a septa (solvent-resistant; coated with Teflon on the solvent side) located on top of the glass collection vials (40 or 60 ml capacity). If required, multiple extractions can be performed per extraction vessel. The arrival and level of solvent in the collection vial is monitored by using an IR sensor. In the event of system failure, an automatic shut-off procedure is instigated. A typical procedure for pressurized fluid extraction is shown in Figure 7.16.

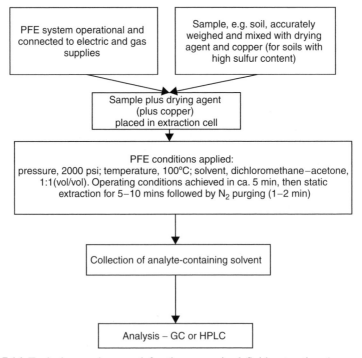

Figure 7.16 Typical procedure used for the pressurized fluid extraction (or accelerated fluid extraction) of solids.

7.7.2 Example 7.6: Pressurized Fluid Extraction of DDT, DDD and DDE from Contaminated Soil[†]

7.7.2.1 Extraction Conditions

These were as follows:

- Sample: 2 g

- PFE conditions: pressure, 2000 psi, temperature, 100°C; static extraction time, 10 min; three static/flush cycles.

Comments Sample placed in stainless-steel extraction cell on top of a filter to prevent cell-frit blockage. 'Hydromatix' was used to fill the headspace in order to reduce solvent consumption.

7.7.2.2 Analysis by GC–MS

Separation and identification of DDT, DDD and DDE was carried out on an HP 5890 Series II Plus gas chromatograph, fitted to an HP 5972A mass spectrometer. A 30 m × 0.25 mm id × 0.25 μm film thickness DB-5ms capillary column was used, with temperature programming from an initial temperature held at 120°C for 2 min before commencing a 5°C min^{-1} rise to 290°C, with a final time of 2 min. The split/splitless injector was held at 280°C and operated in the splitless mode. The mass spectrometer transfer line was maintained at 280°C. Electron impact ionization at 70 eV, with the electron multiplier voltage set at 1500 V, was used, while operating in the single-ion monitoring (SIM) mode.

7.7.2.3 Typical Results

These are shown in Figure 7.17 [6].

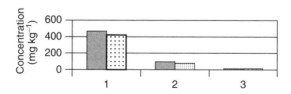

Figure 7.17 Results obtained for the pressurized fluid extraction of DDT, DDD and DDE from contaminated soil, and comparison with those obtained from Soxhlet extraction: ■, Soxhlet; ▦, PFE: 1, DDT; 2, DDD; 3, DDE [6] (cf. DQ 7.13).

[†] Note that the commercial organochlorine pesticide known as 'DDT' is not a single chemical compound. Its major active component (70–80% of the total content) is p,p'-dichlorodiphenyl trichloroethene (p,p'-DDT), along with smaller quantities of the related compounds, p,p'-dichlorodiphenyl dichloroethene (p,p'-DDD) and p,p'-dichlorodiphenyl dichloroethene (p,p'-DDE).

DQ 7.13

Comment on the results obtained in this study (see Figure 7.17).

Answer

It is found that the recoveries of DDT and its metabolites from soil are similar, irrespective of the extraction method being used.

7.8 Matrix Solid-Phase Dispersion

Matrix solid-phase dispersion (MSPD) is analogous to solid-phase extraction (SPE), as described later in Chapter 8. The main difference being that MSPD is used for solid samples. In MSPD, the sample is mixed with an SPE sorbent, e.g. C18 (octadecylsilane (ODS)) from the cartridge. After returning the sorbent-sample mixture to the cartridge, it is eluted as in SPE. The main purpose of the C18 sorbent is to act as an abrasive, thus disrupting the sample's structure, and hence promoting its dispersion within the sorbent and creating a large surface area for solvent interaction. The best ratio of sample to sorbent is 1:4. In practice, it should be possible to adapt the type of sorbent for a particular sample, thereby providing a tailored extraction procedure. A typical scheme for matrix solid-phase dispersion extraction is shown in Figure 7.18.

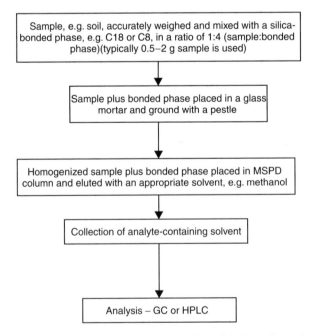

Figure 7.18 Typical procedure used for the matrix solid-phase dispersion of solids.

7.8.1 Example 7.7: Matrix Solid-Phase Dispersion of an Alcohol Ethoxylate (Lutensol, C13 and C15, with an Average Ethoxy Chain of EO7), Spiked onto an Homogenized Fish Tissue

7.8.1.1 Extraction Conditions

These were as follows:

- Sample: 1 g

- MSPD conditions. SPE cartridge washed with 20 ml of hexane, which was then discarded. Elution of the alcohol ethoxylate (AE) was carried out using several solvent systems: (1) 20 ml of methanol followed by 20 ml of acetonitrile, (2) 20 ml of dichloromethane/acetonitrile (1:1 (vol/vol)), followed by 20 ml of methanol, and (3) 20 ml of dichloromethane/acetone (1:1 (vol/vol)). The eluents were taken to 1 ml at 60°C under a gentle stream of nitrogen, and to dryness under nitrogen and in ice.

Comments Sample mixed with 1 ml of methanol, octadecylsilane (ODS, C18) was added (4 g), and the resultant mixture ground until a free-flowing powder was obtained. The mixture was then placed in an empty SPE extraction cartridge with filters at either end.

7.8.1.2 Extract Clean-Up

This was carried out by using two Al–N cartridges. The first of these was conditioned with 20 ml of dichloromethane/methanol (100:5 (vol/vol)) at a flow rate of 2–3 ml min^{-1}. The dried residue from the MSPD step was resolved in 5 ml dichloromethane/methanol (100:5 (vol/vol)), and this was then passed through the Al–N cartridge and collected in a glass vial. The sample container was washed with another 2 × 5 ml of dichloromethane/methanol (100:5 (vol/vol)); these were also passed through the cartridge, which was then washed with a further 5 ml of dichloromethane/methanol (100:5 (vol/vol)). All eluents were collected, taken to 1 ml at 60°C, and to dryness under ice, under a gentle stream of nitrogen.

The second Al–N cartridge was conditioned with 20 ml of cyclohexane at a flow rate of 2–3 ml min^{-1}. Then, the dried extract was resolved in 3 ml of cyclohexane, using a sonic bath. This solution was then transferred to the Al–N cartridge. The vial containing the extract was rinsed with 2 × 5 ml of cyclohexane, and then transferred to the cartridge. The eluate was discarded and the AE eluted with 20 ml of dichloromethane/methanol (100:5 (vol/vol)). This extract was taken to 1 ml at 60°C and to dryness under ice, again under nitrogen.

7.8.1.3 Derivatization Procedure

The AE was derivatized into the corresponding alkyl bromide using HBr fission, by employing the following procedure. The SPE eluent, or standard, in 500 μl of ethyl acetate was dispensed into a 'Chromacol' screw-capped vial. An internal standard (undecanol) was then added. Then, HBr (33% in glacial acetic acid (0.5 ml)) was added to the mixture and the vial was capped tightly. (Note that if acetonitrite (ACN) is present in the solvent, a white precipitate, consistent with CH_3CNHBr, is formed – this does not appear to affect the reaction.) The sample vial ('Chromacol' screw-capped vial) was placed in a heating block, which had been pre-heated to 100–105°C, for 4 h. The vial was removed from the heating block and left to cool to ambient temperature, or in a refrigerator, before opening. The vial was then uncapped and 1,1,1-trichloroethane (500 μl) added. The vial was then recapped and vortex-mixed for 5 s. NaOH (2 M, 4 ml) was then added and the vial vortex-mixed for 10 s, the lower organic layer was allowed to settle, and then removed to a glass vial. The resultant sample was then extracted twice more with 500 μl aliquots of 1,1,1-trichloroethane. The combined extracts were made up to 1.5 ml for analysis.

7.8.1.4 Analysis by GC–MS

Separation and identification of the alcohol ethoxylate was carried out on an HP 5890 Series II Plus gas chromatograph, fitted to an HP 5972A mass spectrometer. A 30 m × 0.25 mm id × 0.25 μm film thickness DB-5ms capillary column was used, with temperature programming from an initial temperature held at 70°C for 2 min before commencing a 10°C min^{-1} rise to 250°C, with a final time of 20 min. The split/splitless injector was held at 230°C and operated in the splitless mode. The mass spectrometer transfer line was maintained at 240°C. Electron impact ionization at 70 eV, with the electron multiplier voltage set at 1500 V, was used, while operating in the single-ion monitoring (SIM) mode. The EI MS detector scan range was 45–425 *m/z*. The key identifier ions used for the alkyl bromide were *m/z* 135 and 137.

7.8.1.5 Typical Results

These are shown in Figure 7.19 [7].

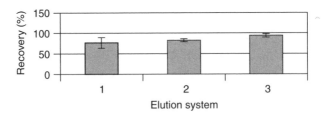

Figure 7.19 Results obtained from the matrix solid-phase dispersion of an alcohol ethoxylate from spiked fish tissue, showing the influence of different elution solvent systems: 1, 20 ml of methanol followed by 20 ml of acetonitrile; 2, 20 ml of di-chloromethane/acetonitrile (1:1 (vol/vol)) followed by 20 ml of methanol; 3, 20 ml of dichloromethane/acetone (1:1 (vol/vol)) [7] (cf. DQ 7.14).

DQ 7.14

Comment on the results obtained in this study (see Figure 7.19).

Answer

It is found that the recovery of alcohol ethoxylate is influenced by the elution solvent. Elution system '3' offers the highest recovery, along with good precision.

7.9 Summary

This chapter has identified the main extraction techniques used for the extraction of organic analytes from solid matrices, e.g. soil. The main purpose of each technique has been to remove the analyte from the matrix as effectively as possible. In some instances, e.g. supercritical fluid extraction, some attempt is made to achieve selectivity of extraction by altering the operating conditions, such as temperature and pressure, or by the addition of an organic modifier. In other situations, the sole purpose of the extraction technique is to remove the analyte from the matrix under the 'strongest' possible conditions without any concern for selectivity. The selectivity in these circumstances results exclusively from the method of analysis which follows. It should also be borne in mind, depending on the level of contamination of the sample, that further extract pre-concentration and/or clean-up may be required prior to analysis in order to achieve trace level analytical results. The methods used for effective pre-concentration are described later in Chapter 10.

While an attempt has been made to place some context on each extraction technique, it should also be noted that most of the techniques described have been validated by the United States Environmental Protection Agency (USEPA) in terms of standard methods. The current USEPA methods for the extraction of pollutants from solid matrices are shown in Table 7.3. It should be noted that the supercritical fluid extraction technique has three specific methods for a range of compounds of environmental importance, whereas other techniques have more general, non-specific methods.

SAQ 7.3

Based on knowledge of the extraction techniques discussed in this chapter, make a comparison of each technique. As a suggestion, the following headings could be used for comparison purposes: brief description of technique; sample mass; extraction time per sample; solvent type; solvent consumption; sequential or simultaneous extraction; relative cost of equipment; level of automation; approval of methods (USEPA).

Table 7.3 Current USEPA methods for the extraction of organic pollutants from solid environmental samples

USEPA method	Extraction technique	Application	Further comments
3540	Soxhlet extraction	Semi-volatile and non-volatile organics from soils, relatively dry sludges and solid wastes	Considered a rugged extraction method because it has very few variables that can adversely affect extraction recovery
3541	Automated Soxhlet extraction	Polychlorinated biphenyls, organochlorine pesticides and semi-volatiles from soils, relatively dry sludges and solid wastes	Allows equivalent extraction efficiency to Soxhlet extraction in 2 h
3545	Accelerated solvent extraction (pressurized fluid extraction)	Semi-volatiles and non-volatile organics from soils, relatively dry sludges and solid wastes	Extraction under pressure using small volumes of organic solvent
3550	Ultrasonic extraction	Semi-volatiles and non-volatile organics from soils, relatively dry sludges and solid wastes	Considered to be less efficient than the other methods detailed; rapid method that uses relatively large solvent volumes. Not appropriate for use with organophosphorus compounds as it may cause destruction of the target analytes during extraction
3560	Supercritical fluid extraction	Semi-volatile petroleum hydrocarbons from soils, relatively dry sludges and solid wastes	Normally uses pressurized CO_2 with additional small volumes of organic solvent. Limited applicability. Relatively rapid extractions
3561	Supercritical fluid extraction	Polycyclic (polynuclear) aromatic hydrocarbons from soils, relatively dry sludges and solid wastes	As method 3560
3562	Supercritical fluid extraction	Polychlorinated biphenyls and organochlorine pesticides from soils, sediments, fly ash, solid-phase extraction media, and other solid materials which are amenable to extraction with conventional solvents	As method 3560

References

1. Saim, N., Dean, J. R., Abdullah, Md. P. and Zakaria, Z., *J. Chromatogr., A*, **791**, 361–366 (1997).
2. Hancock, P. and Dean, J. R., *Anal. Commun.*, **34**, 377–379 (1997).
3. de la Tour, C., *Ann. Chim. Phys.*, **21**, 127–132, 178–182 (1822).
4. Dean, J. R., Barnabas, I. J. and Owen, S. P., *Analyst*, **121**, 465–468 (1996).
5. Richter, B. E., Jones, B. A., Ezzell, J. L., Porter, N. L., Avdalovic, N. and Pohl, C., *Anal. Chem.*, **68**, 1033–1039 (1996).
6. Fitzpatrick, L. J., Dean, J. R., Comber, M. H. I., Harradine, K., Evans, K. P. and Pearson, S., *J. Chromatogr., A*, **874**, 257–264 (2000).
7. Heslop, C. A., 'Identification and extraction of alcohol ethoxylated non-ionic surfactants in environmental samples', *PhD Thesis*, Northumbria University, Newcastle, UK, 2000.

Chapter 8
Liquids

Learning Objectives

- To understand the need for separation and/or pre-concentration for organic compounds in solution.
- To understand the theory of liquid–liquid extraction.
- To be able to carry out solvent extraction in a safe and controlled manner.
- To understand the requirements for solid-phase extraction (SPE).
- To be able to carry out SPE in a safe and controlled manner.
- To understand the theory of solid-phase microextraction (SPME).
- To understand the requirements for SPME.
- To be able to carry out SPME in a safe and controlled manner.

8.1 Liquid–Liquid Extraction

The principal of liquid–liquid extraction (LLE) is that the sample is distributed or partitioned between two immiscible solvents in which the analyte and matrix have different solubilities. The main advantage of this approach is the wide availability of pure solvents and the use of low-cost apparatus. For the background theory on LLE, see Box 6.1 earlier.

DQ 8.1

Can you think of any circumstances when liquid–liquid extraction would be useful?

Answer

Liquid–liquid extraction is required when the analyte is present at a low concentration in a water sample, e.g. river water. In this case, LLE is used to pre-concentrate the analyte from a large volume of water into a small sample volume. In addition, or at the same time, LLE can be used to clean-up the analyte from its matrix. Liquid–liquid extraction is therefore a very useful technique in trace analysis.

8.2 Solvent Extraction

Two common approaches are possible here. In the first approach, the extraction is carried out *discontinuously* where equilibrium is established between two immiscible phases, or secondly by *continuous* extraction. In the case of the latter, equilibrium may not be reached. The selectivity and efficiency of the extraction process is critically governed by the choice of the two immiscible solvents. Using aqueous and organic (e.g. dichloromethane, chloroform, ethylene acetate, toluene, etc.) pairs of solvents, the more hydrophobic analytes prefer the organic solvent while the more hydrophilic compounds prefer the aqueous phase. The more desirable approach is quite often reflected in the nature of the target analyte. For example, if the method of separation to be used is reversed-phase high performance liquid chromatography (HPLC), then the target analyte is best isolated in the aqueous phase. In this situation, the target analyte can then be injected directly into the HPLC system. [Note – the target analyte may well require additional pre-concentration, e.g. solid-phase extraction or solvent elimination (see later), to achieve the appropriate level of sensitivity.] In contrast, if the target analyte is to be analysed by gas chromatography, it is best isolated in organic solvent. In addition, isolation of the target analyte in the organic phase allows solvent evaporation to be employed (see Chapter 10), thus allowing concentration of the target analyte.

The equilibrium process can be influenced by several factors which include adjustment of pH to prevent ionization of acids or bases, by the formation of ion-pairs with ionizable analytes, by the formation of hydrophobic complexes with metal ions, or by adding neutral salts to the aqueous phase to reduce the solubility of the analyte (also known as 'salting out').

In discontinuous extraction, the most common approach uses a separating funnel (Figure 8.1). In this case, the aqueous sample (1 l, at a specified pH) is introduced into a large separating funnel (2 l capacity with a 'Teflon' stopcock) and then 60 ml of a suitable organic solvent, e.g. dichloromethane, is added. The vessel is then sealed with a stopper, and shaken vigorously, either manually or mechanically, for 1–2 min. This shaking process allows thorough interspersion between the two immiscible solvents, thereby maximizing the contact between the two solvent phases and hence assisting mass transfer, and thus allowing efficient partitioning to occur. It is necessary to periodically vent the excess pressure generated during this shaking process.

DQ 8.2
What do you think would happen if you did not release the pressure in the separating funnel?

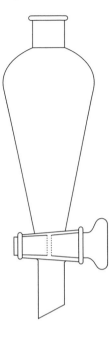

Figure 8.1 A separating funnel used for (discontinuous) liquid–liquid extraction. From Dean, J. R., *Extraction Methods for Environmental Analysis*, Copyright 1998. © John Wiley & Sons Limited. Reproduced with permission.

Answer

The simplest effect would be to 'pop off' the stopper (the path of least resistance). More drastic effects could result if the stopper could not be removed by the excess gas built up inside the separating funnel. This is why in the laboratory it is necessary to wear both a laboratory coat and safety glasses to guard against the unexpected.

After a suitable resting period (10 min), the organic solvent is run off and retained in a collection flask. Fresh organic solvent is then added to the separating funnel and the process repeated again. This should be carried out at least three times in total. The three organic extracts should be combined, either ready for direct analysis or pre-concentration (see Chapter 10), with the exact requirement depending upon the level of contamination.

In some cases the kinetics of the extraction can be slow, such that the equilibrium of the analyte between the aqueous and organic phases is poor, i.e. K_d is very small (see Box 6.1 earlier); if the sample is large, then continuous liquid–liquid extraction can be used. In this situation, fresh organic solvent is boiled, condensed and allocated to percolate repetitively through the analyte-containing aqueous sample. Two common versions of continuous liquid extractors are available, using either lighter-than or heavier-than-water organic solvents (Figure 8.2). Extractions usually take several hours, but do provide concentration factors up to 10^5 (an essential requirement for trace analysis). Obviously, several systems

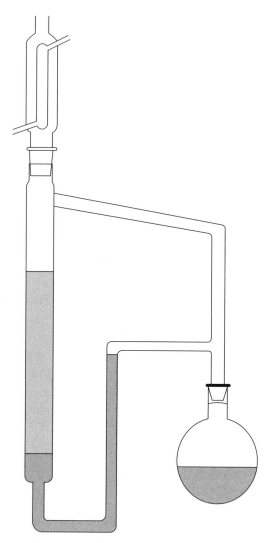

Figure 8.2 A typical system used for continuous liquid–liquid extraction, employing an organic solvent which is heavier than water. From Dean, J. R., *Extraction Methods for Environmental Analysis*, Copyright 1998. © John Wiley & Sons Limited. Reproduced with permission.

can be operated unattended and in series, thus allowing multiple samples to be extracted. Typically, a 1 l sample, pH-adjusted if necessary, is added to the continuous extractor. Then, organic solvent, e.g. dichloromethane (in the case of a system in which the solvent has a greater density than the sample), of volume 300–500 ml is added to the distilling flask together with several boiling chips. The solvent is then boiled, in this case with a water bath, and the extraction process allowed to occur for between 18–24 h. After completion of the extraction process, and sufficient cooling time, the boiling flask is detached and solvent evaporation can then occur (see Chapter 10).

Unfortunately, as with most things, liquid–liquid extraction can suffer from some problems which affect its efficiency.

DQ 8.3

A common problem in LLE is the formation of emulsions, particularly for samples that contain surfactants or fatty materials. What do you think can be done to reduce or eliminate this problem?

Answer

The emulsion can often be broken up by using one of the following approaches, i.e. centrifugation, filtration through a glass wool plug, refrigeration, salting out, or the addition of a small amount of a different organic solvent.

In addition, the rate of extraction may be different for the same analyte depending on the nature of the sample matrix. Obviously, as in all analyses the problem of controlling the level of contamination is crucial. It is essential to use the highest purity solvents (as any subsequent concentration may also concentrate the impurity as well as the analyte of interest) and to wash all associated glassware thoroughly. As well as contamination, care should also be exercised to minimize analyte losses due to adsorption on glass containers. A typical procedure for LLE is described in Figure 8.3.

8.2.1 Example 8.1: Liquid–Liquid Extraction of various Polycyclic Aromatic Hydrocarbons from Water

8.2.1.1 Extraction Conditions:
These were as follows:

- Sample volume: 25 ml
- LLE conditions: sample extracted with 2×3 ml of dichloromethane plus 1 g of salt (NaCl)
- Extracts shaken for 5 min each

Comments Combined extracts placed in a volumetric flask, and internal standard added, prior to analysis.

8.2.1.2 Analysis by GC–MS
Separation and identification of the individual PAHs was carried out on an HP 5890 Series II gas chromatograph, fitted to an HP 5971A mass spectrometer.

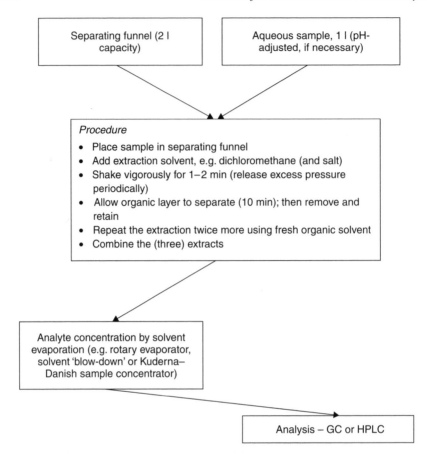

Figure 8.3 Typical procedure employed for liquid–liquid extraction, using a separating funnel.

A 30 m × 0.25 mm id × 0.25 μm film thickness HP-5ms capillary column was used, with temperature programming from an initial temperature held at 90°C for 2 min before commencing a 7°C min^{-1} rise to 285°C, with a final time of 20 min. The split/splitless injector was held at 280°C and operated in the splitless mode, with the split valve closed for 1 min following sample injection. The split flow was set at 40 ml min^{-1}, and the mass spectrometer transfer line was maintained at 280°C. Electron impact ionization at 70 eV, with the electron multiplier voltage set at 1500 V, was used, while operating in the single-ion monitoring (SIM) mode.

8.2.1.3 Typical Results

These are shown in Figure 8.4 [1].

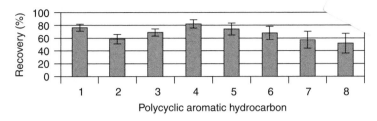

Figure 8.4 Results obtained from the liquid–liquid extraction of various polycyclic aromatic hydrocarbons from water: 1, naphthalene; 2, acenaphthylene; 3, acenaphthene; 4, fluorene; 5, phenanthrene; 6, anthracene; 7, fluoranthene; 8, pyrene [1] (cf. DQ 8.4).

DQ 8.4

Comment on the results obtained in this study (see Figure 8.4).

Answer

*It can be seen that the maximum recovery does not exceed approximately 80%, with a minimum recovery of approximately 55%. The precision of the method, based on three replicate samples, is adequate and reflects the **manual** nature of the LLE process used in this analysis.*

8.3 Solid-Phase Extraction

Solid-phase extraction (SPE), which is sometimes referred to as *liquid–solid extraction*, involves bringing a liquid or gaseous sample into contact with a solid phase or sorbent whereby the analyte is selectively adsorbed onto the surface of the solid phase. The latter is then separated from the solution and other solvents (liquids or gases) are added. The first such solvent is usually a wash to remove possible adsorbed matrix components; eventually, an eluting solvent is brought into contact with the sorbent to selectively desorb the analyte. The focus of this present chapter will be on SPE involving liquid samples and solvents. The solid-phase sorbent is usually packed into small tubes or cartridges, with the system resembling a small liquid chromatography column. The sorbent is also available in round, flat sheets which can be mounted in a filtration apparatus (Empore discs™). In this case, the sorbent resembles that of the commonly used filter paper. Whichever design is used, the sample-containing solvent is forced by pressure or vacuum through the sorbent. By careful selection of the sorbent, the analyte should be retained by the sorbent in preference to other extraneous material present in the sample. This extraneous material can be washed from the sorbent by the passing of an appropriate solvent. Subsequently, the analyte of interest can then be eluted from the sorbent by using a suitable solvent. This solvent is then collected for analysis. Obviously, further sample clean-up or pre-concentration can be carried out, if desired.

DQ 8.5

From this brief introduction, what would you consider to be the important parameters for SPE?

Answer

The choice of sorbent and the solvent system used are of paramount importance for effective pre-concentration and/or clean-up of the analyte in the sample.

The process of SPE should allow more affective detection and identification of the analyte.

8.3.1 *Types of SPE Media*

Generally, SPE sorbents can be divided into three classes, i.e. normal phase, reversed phase and ion-exchange. The most common sorbents are based on silica particles (irregular shaped particles with a particle diameter between 30 and 60 μm) to which functional groups are bonded to surface silanol groups to alter their retentive properties (it should also be noted that unmodified silica is sometimes used). The bonding of the functional groups is not always complete, so some unreacted silanol groups remain. These unreacted sites are polar, acidic sites and can make the interactions with the analytes more complex. In order to reduce the occurrence of these polar sites, some SPE media are 'end-capped', that is, a further reaction is carried out on the residual silanols using a short-chain alkyl group. End-capping is not totally effective. It is the nature of the functional groups which determines the classification of the sorbent. In addition to silica, some other common sorbents are based on florisil, alumina and macroreticular polymers.

Normal-phase sorbents have polar functional groups, e.g. cyano, amino and diol (also included in this category is unmodified silica). The polar nature of these sorbents means that it is more likely that polar compounds will be retained.

DQ 8.6

Suggest some typical functional groups that would be characteristic of polar compounds.

Answer

Typical functional groups which are capable of polar interactions include hydroxyls, amines, carbonyls and groups containing heteroatoms such as oxygen, chlorine, nitrogen, sulfur and phosphorus. Common polar environmental compounds include organochlorine pesticides, organophosphorus pesticides and polychlorinated biphenyls.

In contrast, reversed-phase sorbents have non-polar functional groups, e.g. octadecyl, octyl and methyl, and conversely are more likely to retain non-polar compounds, e.g. polycyclic aromatic hydrocarbons. Ion-exchange sorbents have either cationic or anionic functional groups and when in the ionized form attract compounds of the opposite charge. A cation-exchange phase, such as benzenesulfonic acid, will extract analytes with positive charges (e.g. phenoxyacid herbicides) and vice versa. A summary of the commercially available silica-bonded sorbents is given in Table 8.1.

8.3.2 Cartridge or Disc Format

The design of the SPE device can vary, with each design having its own advantages related to the number of samples to be processed and the nature of the sample and its volume. The most common arrangement is the syringe barrel or cartridge. The cartridge itself is usually made of polypropylene (although glass and polytetrafluorethylene (PTFE) are also available) with a wide entrance, through which the sample is introduced, and a narrow exit (male luer tip). The appropriate sorbent material, ranging in mass from 50 mg to 10 g, is positioned between two frits, at the base (exit) of the cartridge, which act to both retain the sorbent material and to filter out particulate matter. Typically, the frit is made from polyethylene with a 20 μm pore size.

Solvent flow through a single cartridge is typically carried out by using a side-arm flask apparatus (Figure 8.5), whereas multiple cartridges can be simultaneously processed (from 8 to 30 cartridges) by using a commercially available vacuum manifold (Figure 8.6). A variation on this type of cartridge system or syringe filter is when a plunger is inserted into the cartridge barrel. In this situation, the solvent is added to the syringe barrel and forced through the SPE unit by using the plunger. This system is effective if only a few samples are to be processed, for early method development, the SPE method needs to be simple, or if no vacuum system is available.

The most distinctly different approach to SPE is the use of a disc, not unlike a common filter paper. This SPE disc format is referred to by its trade name of Empore™ (discs). The 5–10 μm sorbent particles are intertwined with fine threads of PTFE which results in a disc approximately 0.5 mm thick and a diameter in the range 47 to 70 mm. Empore™ discs are placed in a typical solvent filtration system and a vacuum applied to force the solvent-containing sample through (Figure 8.7). To minimize the dilution effects that can occur, it is necessary to introduce a test-tube into the filter flask to collect the final extract. Manifolds are commercially available for multiple sample extraction using such Empore discs.

Table 8.1 Some commercially available silica-bonded sorbents used in solid-phase extraction

Primary interaction	Phase	Description	Structure
Non-polar	Silica-based	C18, octadecyl	$-Si-C_{18}H_{37}$
	Silica-based	C8, octyl	$-Si-C_8H_{17}$
	Silica-based	C6, hexyl	$-Si-C_6H_{13}$
	Silica-based	C4, butyl	$-Si-C_4H_9$
	Silica-based	C2, ethyl	$-Si-C_2H_5$
	Silica-based	CH, cyclohexyl	—Si—⬡ (cyclohexyl)
	Silica-based	PH, phenyl	—Si—⬡ (phenyl)
	Silica-based	CN, cyanopropyl	$-Si-(CH_2)_3CN$
	Resin-based	ENV$^+$	Hydroxylated polystyrene divinylbenzene
Polar	Silica-based	CN, cyanopropyl	$-Si-(CH_2)_3CN$
	Silica-based	Si, silica	$-Si-OH$
	Silica-based	DIOL, 2,3-dihydroxypropoxypropyl	$-Si-(CH_2)_3-OCH_2CHOHCH_2OH$
	Silica-based	NH$_2$, aminopropyl	$-Si-(CH_2)_3NH_2$
	Silica-based	FL, florisil	$MgO_{3.6}(SiO_2)_{0.1}OH$
	Silica-based	Al, alumina	
Ionic	Silica-based – anion	NH$_2$, aminopropyl	$-Si-(CH_2)_3NH_2$
	Silica-based – anion	SAX, quaternary amine	$-Si-(CH_2)_3N^+(CH_3)_3Cl^-$
	Silica-based – cation	CBA, propylcarboxylic acid	$-Si-(CH_2)_3COOH$
	Silica-based – cation	SCX, benzenesulfonic acid	—Si—⬡—$SO_3^-H^+$
	Silica-based – cation	SCX-2 (PRS), propylsulfonic acid	$-Si-(CH_2)_3SO_3^-H^+$
	Silica-based – cation	SCX-3, ethylbenzenesulfonic acid	—Si—$(CH_2)_2$—⬡—$SO_3^-H^+$

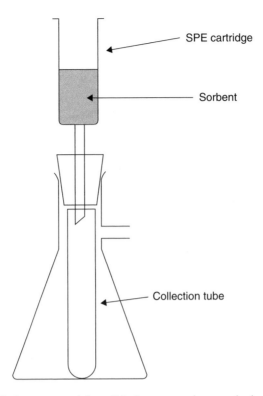

Figure 8.5 A typical system used for solid-phase extraction, employing a cartridge and a side-arm flask apparatus. From Dean, J. R., *Extraction Methods for Environmental Analysis*, Copyright 1998. © John Wiley & Sons Limited. Reproduced with permission.

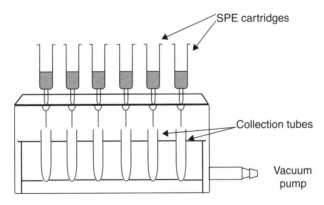

Figure 8.6 Schematic of a commercially available vacuum manifold system used in solid-phase extraction.

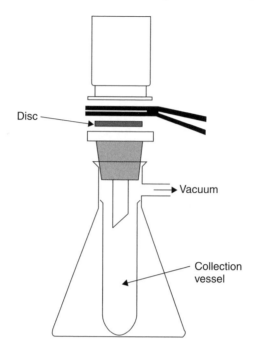

Figure 8.7 A typical system used for solid-phase extraction, employing an 'Empore™' disc and a single side-arm apparatus. From Dean, J. R., *Extraction Methods for Environmental Analysis*, Copyright 1998. © John Wiley & Sons Limited. Reproduced with permission.

Both the cartridge and disc formats have their inherent advantages and limitations. For example, the SPE disc, with its thin sorbent bed and large surface area, allows rapid flow rates of solvent. Typically, one litre of water can be passed through an Empore™ disc in approximately 10 min, whereas with a cartridge system the same volume of water may take approximately 100 min! However, large flow rates can result in poor recovery of the analyte of interest due to there being a shorter time for analyte–sorbent interaction.

8.3.3 Method of SPE Operation

Irrespective of the SPE format, the method of operation is the same and can be divided into five steps (Figure 8.8). Each step is characterized by the nature and type of solvent used, which in turn is dependent upon the characteristics of the sorbent and the sample.

DQ 8.7

What are the five steps for SPE?

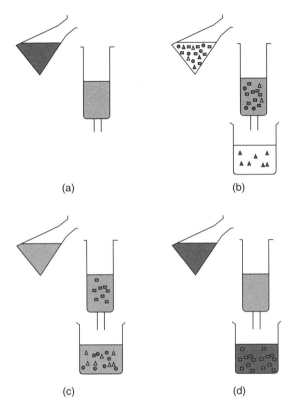

Figure 8.8 Schematic of the method of operation used in solid-phase extraction: (a) sorbent wetted and pre-conditioned; (b) sample applied and retained by sorbent, while some extraneous material (Δ) passes through; (c) remaining extraneous material (Δ) washed off sorbent; (d) analyte (□) eluted from sorbent and collected for analysis.

Answer

The five steps are as follows: wetting the sorbent, conditioning of the sorbent, loading of the sample, rinsing or washing the sorbent to elute extraneous material, and finally elution of the analyte of interest.

Wetting the sorbent allows the bonded alkyl chains, which are twisted and collapsed on the surface of the silica, to be solvated so that they spread open to form a 'bristle'. This ensures good contact between the analyte and the sorbent in the adsorption of the analyte step. It is also important that the sorbent remains wet in the following two steps or otherwise poor recoveries can result. This is followed by *conditioning of the sorbent*, in which solvent or buffer, similar to the test solution that is to be extracted, is 'pulled through' the sorbent. This is followed by *sample loading*, where the sample is forced through the sorbent material by suction, a vacuum manifold or a plunger. By careful choice of the sorbent, it is anticipated that the analyte of interest will be retained by the sorbent in preference to extraneous material and other related compounds of interest that may be present in the sample. Obviously, this ideal situation does not always occur and compounds with similar structures will undoubtedly be retained also. This process is followed by *washing with a suitable solvent which allows unwanted extraneous material to be removed*, without influencing the elution of the analyte of interest. This step is obviously the key to the whole process and is dependent upon the analyte of interest and its interaction with the sorbent material, and the choice of solvent to be used. Finally, *the analyte of interest is eluted from the sorbent* by using the minimum amount of solvent to affect quantitative release. By careful control of the amount of solvent used in the elution step and the sample volume initially introduced onto the sorbent, a pre-concentration of the analyte of interest can be affected. Successful SPE obviously requires careful consideration of the nature of the SPE sorbent, the solvent systems to be used and their influence on the analyte of interest. In addition, it may be that it is not a single analyte that you are seeking to pre-concentrate but a range of analytes. If they have similar chemical structures, then a method can be successfully developed to extract these 'multiple-analytes'. While this method development may seem to be laborious and extremely time-consuming, it should be remembered that multiple-vacuum manifolds are commercially available, as are robotic systems that can carry out the entire SPE process. Once developed, the SPE method can then be used to process large quantities of sample with good precision.

8.3.4 Solvent Selection

The choice of solvent directly influences the retention of the analyte on the sorbent and its subsequent elution, whereas the solvent polarity determines the solvent strength (or ability to elute the analyte from the sorbent in a smaller volume than a weaker solvent). The relative solvent strengths for normal- and reversed-phase sorbents are illustrated in Table 8.2. Obviously, this is the ideal. In some situations, it may be that no individual solvent will perform its function adequately so it is possible to resort to a mixed-solvent system. It should also be noted that for a normal-phase solvent, both solvent polarity and solvent strength are coincident, whereas this is not the case for a reversed-phase sorbent. In practice, however, the solvents normally used for reversed-phase sorbents are restricted to water, methanol, isopropyl alcohol and acetonitrile.

DQ 8.8

What do you think the main effects are for ion-exchange sorbents?

Answer

Solvent strength is not the main effect in this case; pH and ionic strength are the main factors governing analyte retention on the sorbent and its subsequent elution.

Table 8.2 Solvent strengths for normal- and reversed-phase sorbents. From Dean, J. R., *Extraction Methods for Environmental Analysis*, Copyright 1998. © John Wiley & Sons Limited. Reproduced with permission

Solvent strength for normal-phase sorbents		Solvent strength for reversed-phase sorbents
Weakest	Hexane	*Strongest*
	Isooctane	↑
	Toluene	
	Chloroform	
	Dichloromethane	
	Tetrahydrofuran	
	Ethyl ether	
	Ethyl acetate	
	Acetone	
	Acetonitrile	
	Isopropyl alcohol	
	Methanol	
Strongest	Water	*Weakest*

As with the choice of sorbent, some preliminary work is required to affect the best solvents to be used. A typical procedure for SPE is described in Figure 8.9.

8.3.5 Factors Affecting SPE

While the choice of SPE sorbent is highly dependent upon the analyte of interest and the sorbent system to be used, certain other parameters can influence the effectiveness of the SPE methodology. Obviously, the number of active sites available on the sorbent cannot be exceeded by the number of molecules of analyte or otherwise 'breakthrough' will occur. Therefore, it is important to assess the capacity of the SPE cartridge or disc for its intended application. In addition, the flow rate of sample through the sorbent is important; too fast a flow and this will allow minimal time for analyte–sorbent interaction. This must be carefully balanced against the need to pass the entire sample through the cartridge or disc. It is normal, therefore, for an SPE cartridge to operate with a flow rate of $3-10$ ml min^{-1}, whereas rates of $10-100$ ml min^{-1} are typical for the disc format.

Once the analyte of interest has been adsorbed by the sorbent, it may be necessary to wash the sorbent of extraneous matrix components prior to elution of the analyte. Obviously, the choice of solvent is critical in this step, as has been discussed previously. For the elution step, it is important to consider the volume of solvent to be used (as well as its nature).

DQ 8.9
What are the main reasons for the use of SPE for quantitative analysis, for example, HPLC or GC?

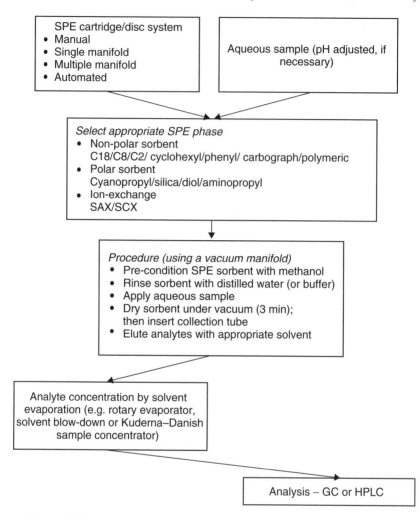

Figure 8.9 Typical procedure used for the solid-phase extraction of liquids.

Answer

The main reasons are (a) pre-concentration of the analyte of interest from a relatively large volume of sample to a small extract volume, and (b) clean-up of the sample matrix to produce a particle-free and chromatographically clean extract.

All of these factors require some method development, either by using a trial-and-error approach or by consultation of the existing literature. It is probable that both are required in practice.

8.3.6 Example 8.2: Solid-Phase Extraction of various Phenols from Water

8.3.6.1 Extraction Conditions

These were as follows:

- Sample volume: 25 ml
- SPE sorbent: PS-DVB, 230 mg
- SPE conditions: conditioning, 5 ml of acetonitrile followed by 5 ml of water; sample loading; interference elution, 2 ml of water; analyte elution, 4 ml of acetonitrile

Comments Sample extract made up to 10 ml with water.

8.3.6.2 Analysis by HPLC

Separation and quantitation was achieved by using a 25 cm × 4.6 mm id ODS2 column with UV detection at 275 nm. The mobile phase was acetonitrile–H_2O– acetic acid (40:59:1), operating under isocratic conditions, at a flow rate of 1 ml min^{-1}. A 100 µl 'Rheodyne' injection loop was used to introduce samples and standards onto the column (at 35°C).

8.3.6.3 Typical Results

These are shown in Figure 8.10 [2].

DQ 8.10

Comment on the results obtained in this study (see Figure 8.10).

Answer

It can be seen that effective pre-concentration of the three different phenols has been demonstrated with adequate precision.

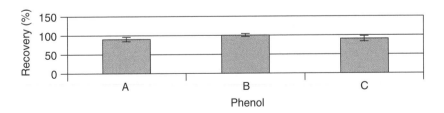

Figure 8.10 Results obtained from the solid-phase extraction of various phenols from river water, at a 'spike level' of 20 ng ml^{-1}: A, phenol; B, 4-nitrophenol; C, 2-methylphenol: calibration range, 0–400 ng ml^{-1}: correlation coefficient(s), 0.9993–0.9979 [2] (cf. DQ 8.10).

SAQ 8.1

Suggest suitable procedures for the SPE of (a) organochlorine pesticides from drinking and surface waters, (b) carbamate pesticides from river water, and (c) acid herbicides from surface water.

8.4 Solid-Phase Microextraction

Solid-phase microextraction (SPME) is the process whereby an analyte is adsorbed onto the surface of a coated-silica fibre as a method of concentration. This is followed by desorption of the analytes into a suitable instrument for separation and quantitation. The theory of SPME is discussed in Box 8.1.

Box 8.1 Theoretical Considerations for SPME

The partitioning of analytes between an aqueous sample and a stationary phase is the main principle of operation of SPME. A mathematical relationship for the dynamics of the absorption process was developed by Louch *et al.* [3]. In this situation, the amount of analyte absorbed by the silica-coated fibre at equilibrium is directly related to its concentration in the sample, as follows:

$$n = K V_2 C_0 V_1 / K V_2 + V_1 \tag{8.1}$$

where n is the number of moles of the analyte absorbed by the stationary phase, K is the partition coefficient of an analyte between the stationary phase and the aqueous phase, C_0 is the initial concentration of analyte in the aqueous phase, V_1 is the volume of the aqueous sample, and V_2 is the volume of the stationary phase.

As was stated earlier, the polymeric stationary phases used for SPME have a high affinity for organic molecules, and hence the values of K are large. These large values of K lead to good pre-concentration of the target analytes in the aqueous sample and a corresponding high sensitivity in terms of the analysis. However, it is unlikely that the values of K are large enough for exhaustive extraction of analytes from the sample. Therefore, SPME is an equilibrium method, but provided that proper calibration strategies are followed it can provide quantitative data.

Louch *et al.* [3] went onto show that in the case where V_1 is very large (i.e. $V_1 \gg K V_2$) the amount of analyte extracted by the stationary phase

Continued on page 159

■ *Continued from page 158* ■
could be simplified to the following:

$$n = K V_2 C_0 \tag{8.2}$$

and hence is not related to the sample volume. This feature can be most effectively exploited in field sampling. In this situation, analytes present in natural waters, e.g. lakes and rivers, can be effectively sampled, pre-concentrated and then transported back to the laboratory for subsequent analysis. The dynamics of extraction are controlled by the mass transport of the analytes from the sample to the stationary phase of the silica-coated fibre. The dynamics of the absorption process have been mathematically modelled [3]. In this work, it was assumed that the extraction process is diffusion-limited. Therefore, the amount of sample absorbed, when plotted as a function of time, can be derived by solving *Fick's Second Law of Diffusion*. A plot of the amount of sample absorbed versus time is termed the extraction profile. The dynamics of extraction can be increased by stirring the aqueous sample.

8.4.1 Experimental

The most common approach for SPME is its use for GC, although its coupling to HPLC has also been reported. The SPME device consists of a fused-silica fibre, coated with a gas chromatographic stationary phase, e.g. polydimethylsiloxane. In addition, various other stationary phases are available for SPME (Table 8.3). The small size and cylindrical geometry allow the fibre to be incorporated into a syringe-type device (Figure 8.11). This allows the SPME device to be effectively used in the normal unmodified injector of a gas chromatograph. As can be seen in Figure 8.11, the fused-silica fibre (approximately 1 cm) is connected to a stainless-steel tube for mechanical strength. This assembly is mounted

Table 8.3 Some commercially available SPME fibre coatings

7 μm polydimethylsiloxane (bonded)
30 μm polydimethylsiloxane (non-bonded)
100 μm polydimethylsiloxane (non-bonded)
85 μm polyacrylate (partially cross-linked)
60 μm polydimethylsiloxane/divinylbenzene (partially cross-linked)
65 μm polydimethylsiloxane/divinylbenzene (partially cross-linked)
75 μm polydimethylsiloxane/Carboxen (partially cross-linked)
65 μm Carbowax/divinylbenzene (partially cross-linked)
50 μm Carbowax/Template resin (partially cross-linked)

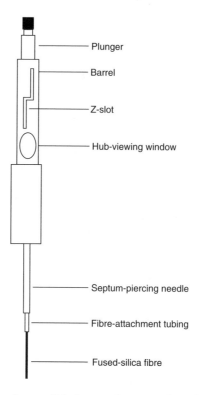

— Plunger

— Barrel

— Z-slot

— Hub-viewing window

— Septum-piercing needle

— Fibre-attachment tubing

— Fused-silica fibre

Figure 8.11 Schematic of a solid-phase microextraction device. Reprinted with permission from Zhang, Z., Yang, M. and Pawliszyn, J., *Anal. Chem.*, **66**, 844A–853A (1994). Copyright (1994) American Chemical Society.

within the syringe barrel for protection when not in use. For SPME, the fibre is withdrawn into the syringe barrel, and then inserted into the sample-containing vial for either solution or air analysis. At this point, the fibre is exposed to the analyte(s) by pressing down the plunger for a pre-specified time. After this pre-determined time-interval, the fibre is withdrawn back into its protective syringe barrel and withdrawn from the sample vial. The SPME device is then inserted into the hot injector of the gas chromatograph and the fibre exposed for a pre-specified time. The heat of the injector desorbs the analyte(s) from the fibre prior to GC separation and detection. SPME can carried out either manually or by an autosampler. As the exposed fibre is an active site for adsorption of not only the analytes of interest but also air-borne contaminants, it is essential that the SPME fibre is placed in the hot injector of the chromatograph prior to adsorption/desorption of the analytes of interest, in order to remove potential interferents.

A typical procedure used for SPME is described in Figure 8.12.

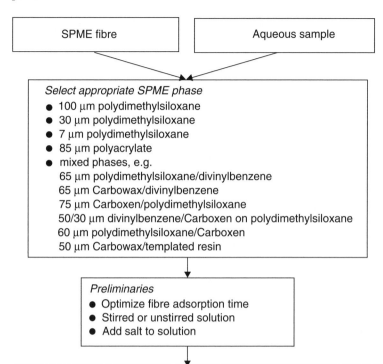

Figure 8.12 Typical procedure used for the solid-phase microextraction of liquids.

8.4.2 Example 8.3: Solid-Phase Microextraction of BTEX from Water[†]

8.4.2.1 Extraction Conditions

These were as follows:

- Sample volume: 10 ml
- Fibre: 100 μm polydimethylsiloxane

Comments SPME fibre inserted into either the sample or headspace above the sample (with/without stirring; with/without salt) for varying amounts of time.

8.4.2.2 Analysis by GC

Separation and identification of the BTEX mixture was carried out on a Carlo Erba HRGC 5300 Mega Series gas chromatograph, with split/splitless injection

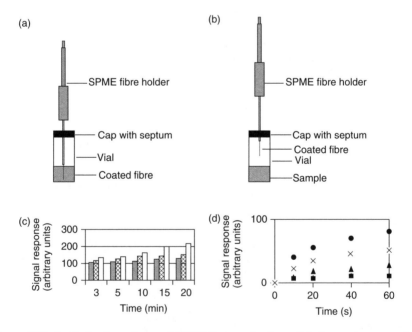

Figure 8.13 Analysis of *o*-xylene and BTEX (in water) using solid-phase microextraction: (a) direct SPME fibre mode; (b) headspace SPME fibre mode; (c) results obtained for *o*-xylene using mode (a); (d) results obtained for BTEX using mode (b): ■, no stirring; ⊠, with stirring; □, with stirring, plus salt: ■, benzene; ♦, toluene; ▲, ethylbenzene; ●, *m*-, *p*-xylene(s); ×, *o*-xylene [4] (cf. DQ 8.11).

[†] The acronymn 'BTEX' is commonly used to refer to a low-boiling-point mixture of benzene, toluene, ethylbenzene and xylene(s).

and flame ionization detection. A 30 m × 0.25 mm id × 0.1 μm film thickness DB-5 capillary-column was used, with temperature programming from an initial temperature held at 50°C for 3 min before commencing a 16°C min^{-1} rise to 120°C, with a final hold time of 7 min. The detector temperature was set at 250°C.

8.4.2.3 Typical results
These are shown in Figure 8.13 [4].

DQ 8.11
Comment on the results obtained in this study (see Figure 8.13).

Answer
In the 'direct mode' (Figures 8.13(a, c)) the SPME fibre has been exposed to o-*xylene for increasing amounts of time in three different sequences. These are unaided ('no stirring'), 'with stirring', and finally 'with stirring and salt'. It is noted that 'no stirring' results in the smallest signal obtained, after GC(FID) analysis, in all cases. Stirring provides a greater signal which can be improved by the addition of salt ('salting out'). What should also be noted is the time-scale (minutes) in the 'direct mode' needed to achieve an appropriate signal response. In the 'headspace mode' (Figures 8.13(b, d)), a range of BTEX compounds have been exposed to the fibre prior to GC(FID) analysis. In this case, it should be observed that (i) each compound results in a different response, (ii) the response increases with respect to time, and (iii) the time-domain is in seconds (**not** minutes, as in the 'direct mode').*

SAQ 8.2
Based on knowledge of the extraction techniques discussed in this chapter, make a comparison of each technique. As a suggestion, the following headings can be used for comparison purposes: brief description of technique; sample volume; extraction time per sample; solvent consumption; relative cost of equipment; 'acceptability'; approval of methods (USEPA).

Summary

The presence of trace organics in natural and waste waters can often cause a problem in terms of the selected analytical technique. In order to be able to quantify the concentrations of trace organics in aqueous samples, appropriate methods of pre-concentration therefore need to be selected. This present chapter has summarized the main methods available for such pre-concentration procedures. The traditional approach has utilized liquid–liquid extraction. However, since the 1970s solid-phase extraction (SPE) has become increasingly popular, particularly as it is possible to automate the procedure. Most recently (the 1990s), the use of solid-phase microextraction (SPME) has offered an alternative approach to pre-concentration. However, it is not foreseen that SPME will replace SPE, but rather offer an alternative method which is 'portable' and hence can be applied outside of the laboratory.

References

1. Palmira Arenaz-Laborda, M., 'Extraction of polycyclic aromatic hydrocarbons from soil using hot water extraction coupled with solid phase microextraction', *MSc Dissertation*, Northumbria University, Newcastle, UK, 1998.
2. Madier, C. 'Extraction of phenols from river water', *BSc Project*, Northumbria University, Newcastle, UK, 1997.
3. Louch, D., Motlagh, S. and Pawliszyn, J., *Anal. Chem.*, **64**, 1187–1199 (1992).
4. Ahmed, H. K., 'Separation of common herbicides from water samples using SPE, followed by HPLC–UV', *MSc Dissertation*, Northumbria University, Newcastle, UK, 1996.
5. Biziuk, M., Namiesnik, J., Czerwinski, J., Gorlo, D., Makuch, B., Janicki, W., Polkowska, Z. and Wolska, L., *J. Chromatogr., A*, **733**, 171–183 (1996).
6. Honing, M., Riu, J., Barcelo, D., van Baar, B. L. M. and Brinkman, U. A. Th., *J. Chromatogr., A*, **733**, 283–294 (1996).
7. Vink, M. and van der Poll, J. M., *J. Chromatogr., A*, **733**, 361–366 (1996).

Chapter 9

Volatile Compounds

Learning Objectives

- To be able to analyse volatile compounds in the atmosphere.
- To understand the requirements for thermal desorption.
- To be able to carry out thermal desorption in a safe and controlled manner.
- To understand the requirements for purge-and-trap extraction.
- To be able to carry out purge-and-trap extraction in a safe and controlled manner.

9.1 Introduction

Volatile compounds in the atmosphere, workplace and on industrial sites need to be monitored with regard to safety considerations, e.g. emissions to the atmosphere, or occupational standards. These volatile compounds, e.g. BTEX, can be trapped either on a solid support material (e.g. for thermal desorption), or liberated from a water sample and then trapped (e.g. via purge-and-trap), prior to analysis.

> **DQ 9.1**
> What does BTEX stand for?
>
> *Answer*
> *This acronym stands for Benzene, Toluene, Ethylbenzene, and ortho-, meta- and para-Xylenes (see also Example 8.3 earlier).*

9.2 Thermal Desorption

Volatile analytes in the atmosphere can be adsorbed onto solid support material. The role of thermal desorption, in conjunction with gas chromatography, is to desorb the analytes from the collection material by the application of heat and then analyse them by chromatography.

DQ 9.2

Which variables do you think will influence the desorption process?

Answer

The desorption process can be influenced by varying the temperature of the collection material and the carrier-gas flow rate.

The technique itself is solvent-less (no organic solvents are required) and can also be automated.

It is obviously important that the sample is heated in a manner which maximizes the recovery of the adsorbed analytes without altering their chemical compositions. In order to maintain analyte integrity, relatively cool temperatures (approximately 100°C) are used. Unfortunately, however, the desorption of analytes at these temperatures may be slow.

DQ 9.3

What consequence will the slow desorption of analytes have when directly interfaced to the GC column?

Answer

, *The analytes may produce broad, poorly resolved peaks in the chromatogram.*

However, this is not always the case, and some analytes may desorb rapidly, producing good peak shapes (compare the situation with SPME (see Chapter 8), whereby the hot injector of the gas chromatograph is used to desorb analytes and introduce them into the instrument). Poor resolution can be improved by trapping the volatile analytes cryogenically onto a GC column before initializing the temperature programme. This can be achieved by utilizing the GC oven's cryogenic function or by installing a cryogenic focuser, which uses either liquid nitrogen or carbon dioxide as a cooling agent, at the head of the column. While this adds additional cost to the analysis, it does provide the capability to control the sampling time and temperatures required for the analysis of the compounds of interest.

In some situations, the volatile compounds under investigation may have been collected onto another tube or cartridge filled with sorbent. These tubes are used to concentrate volatile compounds, e.g. from air samples in the environment or workplace. In this manner, air samples in the atmosphere can be regularly monitored. By placing the inlet of the cartridge at the suspected source of contaminant, a measured volume of air is pulled through the cartridge containing the sorbent by using a vacuum pump. The cartridge is then inserted into the thermal desorption unit where analytes are then desorbed either directly to the gas chromatograph (Figure 9.1), or first to a cryogenic trap and then the chromatograph.

Thermal desorption has been applied to the following common sample types:

- Food samples, e.g. natural aromas and for residuals and contaminants
- Polymer samples, e.g. additives such as plasticizers

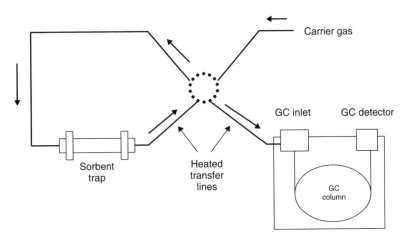

Figure 9.1 Illustration of a typical layout for thermal desorption, where the desorption unit (set in the desorption position) is connected directly to a gas chromatograph: ⟶ indicates the flow of carrier gas.

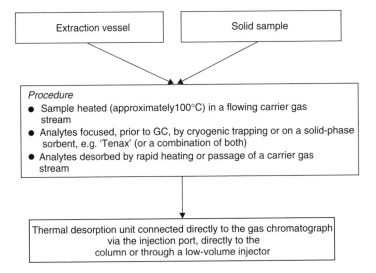

Figure 9.2 A typical procedure used for the thermal desorption of solid samples.

- Environmental samples, e.g. contaminants in soils
- Forensic samples, e.g. firearms residues
- Residual solvents, e.g. in goods manufacturing and dry cleaning
- Air sampling onto sorbent tubes

A typical procedure used for the thermal desorption of solid samples is described in Figure 9.2.

9.3 Purge-and-Trap

Purge-and-trap is widely used for the extraction of volatile organic compounds from aqueous samples: this is then followed by gas chromatography. The method involves the introduction of an aqueous sample (typically 5 ml) into a glass sparging vessel. The sample is then purged (Figure 9.3). This involves passing high-purity nitrogen gas bubbles through the sample to remove the volatile analytes to be trapped. Typical flow rates are 40–50 ml min^{-1}, for 10–12 min. An additional dry purge may be required, where gas only is passed through the trap to remove excess water. The extracted volatile organics are then transferred to a trap, e.g. 'Tenax', silica or charcoal, at ambient temperature. This is followed by the desorption step (Figure 9.4). The trap is heated (180–250°C) and then back-flushed

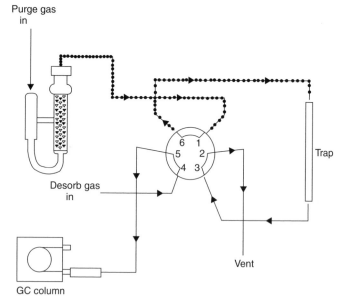

Figure 9.3 Illustration of a typical layout for purge-and-trap extraction of volatile organic compounds from aqueous samples – in 'purge mode': ⟶ indicates sample pathway.

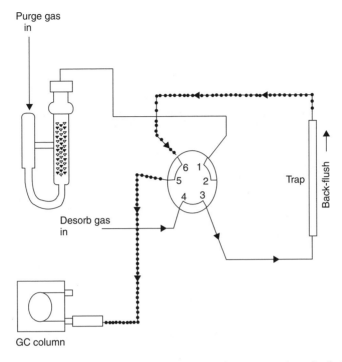

Purge gas
in

Desorb gas
in

Trap

Back-flush

GC column

Figure 9.4 Illustration of a typical layout for purge-and-trap extraction of volatile organic compounds from aqueous samples – in 'desorb mode': ⟶ indicates sample pathway.

with nitrogen to send the sample to the GC column. Typical desorption times are 2–4 min, with nitrogen flow rates of 1–2 ml min^{-1}, for narrow-bore columns. This allows the desorption of volatile organic compounds in a narrow band. The desorbed compounds are transferred via a heated transfer line to the injector of a gas chromatograph for separation and detection. In order to maintain the integrity of the trap, it is periodically cleaned. This is achieved by heating the trap to remove contaminants and residual water. A higher temperature than the desorb temperature is used for ca. 8 min. A typical procedure used for the thermal desorption (purge-and-trap) of aqueous samples is described in Figure 9.5.

9.3.1 Example 9.1: Purge-and-Trap Extraction of BTEX from Water

9.3.1.1 Extraction Conditions

These were as follows:

- Sample volume: 2–10 ml

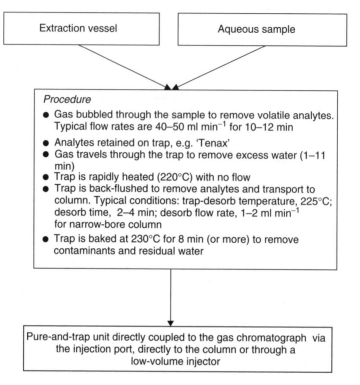

Figure 9.5 A typical procedure used for the purge-and-trap (thermal desorption) extraction of volatile organic compounds from aqueous samples.

- Purge-and-trap conditions: sample sparged for 2–5 min using N_2; BTEX mixture trapped on 'Tenax' trap maintained at 20°C for 1–5 min; analytes desorbed by rapid heating to 260°C for 1 min

Comments GC column initially maintained at 50°C to concentrate analytes.

9.3.1.2 Analysis by GC

Separation and identification of the BTEX mixture was carried out on a Carlo Erba HRGC 5300 Mega Series gas chromatograph, with split/splitless injection and flame ionization detection. A 30 m × 0.25 mm id × 0.1 μm film thickness DB-5 capillary column was used, with temperature programming from an initial temperature held at 50°C for 3 min before commencing a 16°C min⁻¹ rise to 120°C, with a final hold time of 7 min. The detector temperature was set at 250°C.

9.3.1.3 Typical Results

These are shown in Figure 9.6 [1].

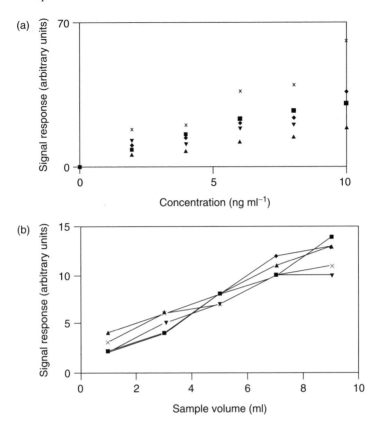

Figure 9.6 Results obtained for the purge-and-trap extraction of BTEX from water, showing (a) the calibration graphs, and (b) the influence of sample volume: ♦, benzene; ■, toluene; ▲, ethylbenzene; ×, *m-*, *p*-xylene; ▼, *o*-xylene [1] (cf. DQ 9.4).

DQ 9.4

Comment on the results obtained in this study (see Figure 9.6).

Answer

Linear calibration graphs (Figure 9.6(a)) are shown for all of the BTEX components in the 0–10 ng ml⁻¹ range. In addition, it is observed (Figure 9.6(b)) that the larger the sample volume, then the larger the signal. This situation is obviously important if you are seeking to carry out trace analysis. For such analysis, more sample is required in order to achieve a lower detection limit.

Summary

The methods described in this chapter are rather specialist in nature and you may not come across them in the undergraduate laboratory. However, their importance cannot be underestimated in the 'real' world where contamination from volatile organic compounds can cause health problems and hence require regular monitoring.

References

1. Leconte, A., 'Comparison of purge-and-trap–GC with headspace solid-phase microextraction–GC for the analysis of BTEX in water', MSc Dissertation, Northumbria University, Newcastle, UK, 1997.

Chapter 10

Pre-Concentration Using Solvent Evaporation

Learning Objectives

- To understand the need for pre-concentration for organic compounds in organic solvents.
- To be able to carry out rotary evaporation in a safe and controlled manner.
- To understand the requirements for Kuderna–Danish evaporative concentration.
- To understand the requirements for an automated evaporative concentration system.
- To be able to carry out gas 'blow-down' in a safe and controlled manner.

10.1 Introduction

Pre-concentration is concerned with the reduction of a larger sample into a smaller sample size. It is most commonly carried out by using solvent evaporation procedures after an extraction technique (see, for example, Chapters 7 and 8). The most common approaches for solvent evaporation are rotary evaporation, Kuderna–Danish evaporative concentration, the automated evaporative concentration system (EVACS) or gas 'blow-down'. In all cases, the evaporation method is slow, with a high risk of contamination from the solvent, glassware and blow-down gas.

DQ 10.1

When would solvent evaporation be required?

Answer

Solvent evaporation would be used when you are trying to perform ultra-trace analysis. It allows a sample extract to be further pre-concentrated. An example would be analysis of pesticides in natural water. Initially, solid-phase extraction may have been performed to concentrate the pesticides from the water. However, to detect pesticides in natural waters will require additional pre-concentration via solvent-evaporation procedures.

10.2 Rotary Evaporation

In this method, the solvent is removed under reduced pressure by mechanically rotating the flask containing the sample in a controlled-temperature water bath (Figure 10.1). The (waste) solvent is condensed and collected for disposal.

DQ 10.2

What type of problems might you encounter with this approach?

Answer

Problems can occur due to loss of volatile analytes, adsorption onto glassware, entrainment of analyte in the solvent vapour and the uncontrollable overall evaporation process.

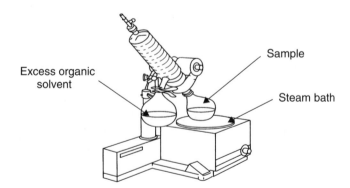

Figure 10.1 A typical rotary evaporation system used for the pre-concentration of compounds in organic solvents.

10.3 Kuderna–Danish Evaporative Concentration

The Kuderna–Danish evaporative condenser was developed in the laboratories of Julius Hyman and Company, Denver, Colorado, USA [1, 2]. This consists of an evaporation flask (500 ml) connected at one end to a Snyder column and the other end to a concentrator tube (10 ml) (Figure 10.2). The sample-containing organic solvent (200–300 ml) is placed in the apparatus, together with one or two boiling chips, and heated with a water bath. The temperature of the water bath should be maintained at 15–20°C above the boiling point of the organic solvent. The positioning of the apparatus should allow partial immersion of the concentrator tube in the water bath but also allow the entire lower part of the evaporation flask to be bathed with hot vapour (steam). Solvent vapours then rise and condense within the Snyder column. Each section of this column consists of a narrow opening covered by a loose-fitting glass insert. Sufficient pressure needs to be generated by the solvent vapours to force their way through the column. Initially, a large amount of condensation of these vapours returns to the

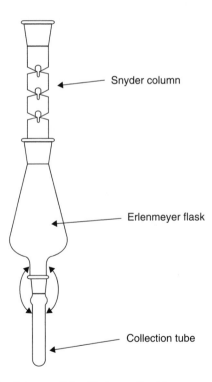

Snyder column

Erlenmeyer flask

Collection tube

Figure 10.2 Schematic diagram of the Kuderna–Danish evaporative concentration condenser system. From Dean, J. R., *Extraction Methods for Environmental Analysis*, Copyright 1998. © John Wiley & Sons Limited. Reproduced with permission.

bottom of the Kuderna–Danish apparatus. In addition to continually washing the organics from the sides of the evaporation flask, the returning condensate also contacts the rising vapours and assists in the process of recondensing volatile organics. This process of solvent distillation concentrates the sample to approximately 1–3 ml in 10–20 min. Escaping solvent vapours are recovered by using a condenser and collection device. The major disadvantage of this method is that violent solvent eruptions can occur in the apparatus, thus leading to sample losses. 'Micro-Snyder' column systems can be used to reduce the solvent volume still further.

10.4 Automated Evaporative Concentration System

In this system, solvent from a pressure-equalized reservoir (500 ml capacity) is introduced, under controlled flow, into a concentration chamber (Figure 10.3) [3]. Glass indentations regulate the boiling of solvent so that bumping does not occur. This reservoir is surrounded by a heater. The solvent reservoir inlet is situated under the level of the heater, just above the final concentration chamber. This chamber is calibrated to 1.0 and 0.5 ml volumes. A distillation column is connected to the concentration chamber. Located near the top of the column are four rows of glass indentations which serve to increase the surface area. Attached to the top of the column is a solvent-recovery condenser with an outlet to collect, and hence recover, the solvent.

To start a sample run, the apparatus is operated with 50 ml of high-purity solvent under steady uniform conditions at total reflux for 30 min to bring the system to equilibrium. Then, the sample is introduced into the large reservoir, either as a single volume or over several time-intervals. (NOTE – A boiling point difference of approximately 50°C is required between solvent and analyte for the highest recoveries.) The temperature is maintained to allow controlled evaporation. For semi-volatile analytes, this is typically at 5°C higher than the boiling point of the solvent. The distillate is withdrawn while keeping the reflux ratio as high as possible. During operation, a sensor monitors the level of liquid, thus allowing the heating to be switched off or on automatically (when liquid is present, the heat is on, and vice versa). After evaporation of the sample below the sensor level, the heating is switched off. After 10 min, the nitrogen flow is started for the final concentration stage from 10 ml to 1 ml (or less). Mild heat can be applied according to the sensitivity of the solvent and analyte to thermal decomposition. When the liquid level drops below the tube, stripping almost stops. The tube is sealed at the bottom, so that the nitrogen is dispersed above the sample and the reduction of volume becomes extremely slow. This prevents the sample from going to dryness, even if left for hours. The sample is drained and the column is rinsed with two 0.5 ml aliquots of solvent. Further concentration can take place, if required.

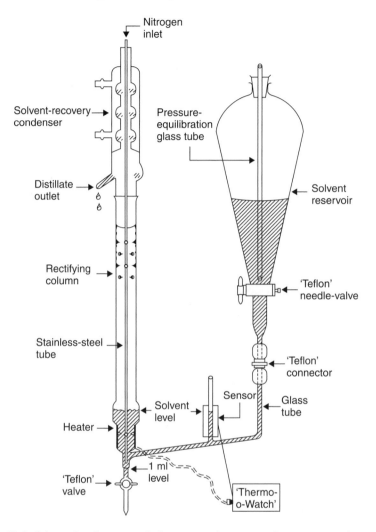

Figure 10.3 Schematic diagram of the automatic evaporative concentration system (EVACS): ▨, solvent; ☐, vapour. Reprinted with permission from Ibrahim, E. A., Suffet, I. H. and Sakla, A. B., *Anal. Chem.*, **59**, 2091–2098 (1987). Copyright (1987) American Chemical Society.

10.5 Gas 'Blow-Down'

In this method, a gentle stream of (high-purity) purge gas is passed over the surface of the extract (Figure 10.4). The latter may be contained inside a conical-tipped or similar vessel. In this situation, the purge gas is directed towards the side of the vessel, and not directly onto the top of the extract, in order to induce

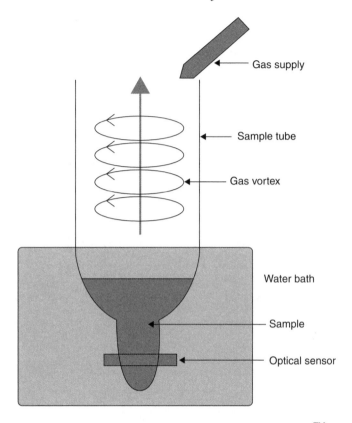

Figure 10.4 Schematic of a typical gas 'blow-down' system (Turbovap™) system used for the pre-concentration of compounds in organic solvents.

a swirling action. The extract-containing vessel may be partially immersed in a water bath to speed up the evaporation process.

DQ 10.3

How else might the evaporation rate be increased?

Answer

Alternative ways to speed up the evaporation process are to increase the flow rate of the impinging gas (NOTE – too high a rate and losses may occur), alter its position with respect to the extract surface, or increase the solvent extract surface area available for evaporation.

If carry-over of trace quantities of the aqueous sample or high-boiling point solvent has occurred, it may not be possible to evaporate them without significant losses of the analyte of interest. The extract may be taken to dryness or left as a small volume (~1 ml).

A comparison between the EVACS and Kuderna–Danish (K–D) approaches for solvent concentration during environmental analysis of trace organic chemicals has been made [3]. Some selected results for a range of organic compounds, lower-boiling-point compounds, and volatile organic compounds are shown in Tables 10.1–10.3, respectively. The authors of this study [3] conclude that the evaporative concentration system (EVACS) is more efficient than the Kuderna–Danish evaporation technique for concentrating semi-volatile–volatile organic analytes (bpt 80–385°C) from a low-boiling solvent (e.g. dichloromethane, bpt 40°C). The advantages of the EVACS over the K–D approach were described as follows:

- The temperature control associated with the EVACS could allow evaporation of a wider range of solvents and avoid violent boiling (bumping) which often happens with the K–D technique.

- Fine adjustment of the temperature gives better control on compounds that are sensitive to thermal decomposition, e.g. 2,4-dichlorophenol.

- The time of operation is known for the EVACS, depending on the evaporation rate needed for a certain application. This is not possible with the K–D

Table 10.1 Comparative recoveries for two-step K–D and EVACS approaches for a range of organic compounds[a] (from 200 to 1 ml) [3]. Reprinted with permission from Ibrahim, E. A., Suffet, I. H. and Sakla, A. B., *Anal. Chem.*, **59**, 2091–2098 (1987). Copyright (1987) American Chemical Society

Compound	Initial concentration (mg l^{-1})	Two-step K–D (% recovery)[b]	EVACS (% recovery)[b]
Acetophenone	0.241	82($\pm$1)	93($\pm$6)[c]
Isophorone	0.243	85($\pm$3)	88($\pm$4)
2,4-Dichlorophenol	0.238	78($\pm$7)	99($\pm$2)[c]
Quinoline	0.201	81($\pm$12)	100($\pm$3)[c]
1-Chlorodecane	0.239	93($\pm$2)	91($\pm$6)[c]
2-Methylnaphthalene	0.174	87($\pm$1)	98($\pm$2)[c]
Biphenyl	0.226	83($\pm$2)	92($\pm$6)[c]
1-Chlorododecane	0.239	87($\pm$4)	96($\pm$2)[c]
Diacetone-L-sorbose	0.214	—	98($\pm$4)
Anthracene	0.159	93($\pm$5)	97($\pm$4)
Dioctylphthalate	0.211	71($\pm$4)	99($\pm$4)[c]

[a]Conditions: solvent, dichloromethane; evaporation rate, 4 ml min^{-1} for EVACS, although uncontrollable for K–D; N$_2$ flow rate, 1 ml s^{-1}; number of determinations, 4.
[b]Figures in parentheses indicate standard deviations (SDs).
[c]Significantly better recovery values determined statistically by analysis of variance (ANOVA) calculations, at $t_{0.975}$ (theoretical significance).

Table 10.2 Comparative recoveries for the two-step K–D and EVACS approaches for a range of lower-boiling-point compounds[a] (from 200 to 1 ml) [3]. Reprinted with permission from Ibrahim, E. A., Suffet, I. H. and Sakla, A. B., *Anal. Chem.*, **59**, 2091–2098 (1987). Copyright (1987) American Chemical Society

Compound	Initial concentration (mg l^{-1})	Two-step K–D (% recovery)[b]	EVACS (% recovery)[b]
Ethyl butyrate	0.27	70($\pm$2)	70($\pm$3)
Ethylbenzene	0.29	72($\pm$2)	71($\pm$3)
Cyclohexanone	0.29	74($\pm$1)	76($\pm$4)
Anisole	0.29	72($\pm$1)	76($\pm$3)[c]
1,4-Dichlorobenzene	0.27	75($\pm$2)	80($\pm$2)[c]
2-Ethylhexanol	0.26	76($\pm$1)	82($\pm$3)[c]
Tolunitrile	0.27	82($\pm$2)	80($\pm$1)
Naphthalene	0.26	88($\pm$4)	96($\pm$3)[c]
Benzothiazole	0.27	100($\pm$1)	99($\pm$3)
Ethyl cinnamate	0.26	90($\pm$1)	98($\pm$6)

[a]Conditions: solvent, dichloromethane; evaporation rate, 4 ml min^{-1} for EVACS, although uncontrollable for K–D; N$_2$ flow rate, 1 ml s^{-1}; number of determinations, 4.
[b]Figures in parentheses indicate standard deviations (SDs).
[c]Significantly better recovery determined statistically by analysis of variance (ANOVA) calculations, at $t_{0.975}$ (theoretical significance).

Table 10.3 Comparative recoveries for the two-step K–D and EVACS approaches for a range of volatile organic compounds[a] (from 200 to 1 ml) [3]. Reprinted with permission from Ibrahim, E. A., Suffet, I. H. and Sakla, A. B., *Anal. Chem.*, **59**, 2091–2098 (1987). Copyright (1987) American Chemical Society

Compound	Boiling point (°C)	Initial concentration (mg l^{-1})	Two-step K–D (% recovery)[b]	EVACS (% recovery)[b,c]
Trichloroethylene	87.0	0.40	17($\pm$3)	24($\pm$1)
Benzene	80.1	0.52	29($\pm$2)	38($\pm$1)
Perchloroethylene	121.14	0.40	53($\pm$3)	63($\pm$4)
Toluene	110.6	0.29	53($\pm$1)	68($\pm$6)
Chlorobenzene	132.22	0.41	55($\pm$3)	68($\pm$1)

[a]Conditions: solvent, dichloromethane; evaporation rate, 2 ml min^{-1} for EVACS, although uncontrollable for K–D; N$_2$ flow rate, 1 ml s^{-1}; number of determinations, 3.
[b]Figures in parentheses indicate standard deviations (SDs).
[c]Significantly better recovery determined statistically by analysis of variance (ANOVA) calculations, at $t_{0.975}$ (theoretical significance).

technique, where vigorous boiling occurs at the beginning, followed by a cooling-down during operation, depending on the surrounding conditions, due to a lack of control of the heat supplied.

- Any volume of sample can be concentrated in the same container with continuous operation down to a 0.5–1 ml final volume.

- Sample contamination via transfer of the K–D glassware is eliminated.

- The EVACS avoids the risk of loosing samples by evaporation to dryness even if the system is left unattended for hours because of the special design of the nitrogen tube. Evaporation to dryness can happen with the K–D system during the first or second step.

- The apparatus is easy to build and operate.

- Automation makes the process very easy, but even without automation, the apparatus needs only minimal attention to maintain the same level in the concentration chamber.

- Economically, the EVACS has the advantage that solvents can be recovered for re-use for similar samples, especially when evaporating large volumes of solvents, without disturbing the process.

In addition, the EVACS procedure has been designed to avoid most of the difficulties and weaknesses associated with the conventional K–D method. Possible errors that are still associated with both of the EVACS and K–D techniques can result from the following:

- Cross-contamination can occur between samples – thus careful cleaning is required between each sample stage.

- The possibility of thermal decomposition of certain compounds, which therefore requires fine adjustment of the heat supply, especially during nitrogen stripping.

- Quantitative errors can result from the manual addition of internal standards into the final volume prior to analysis.

- Instrumental analytical errors – these can be minimized by the addition of internal standards.

Summary

In order to achieve the lowest levels of any analyte in the environment, we require the use of appropriate samples in which the analyte is homogeneously distributed. Then, by careful sample pre-treatment and selection of the most appropriate extraction/digestion procedures, we can then go ahead and carry out the final analysis. Now, assuming that we have taken due regard to minimize the risk of contamination, we should then expect the analytical instrument to record an appropriate signal (response). However, this may not always be the case. Often, the concentration levels of potential environmental contaminants are so low that an extra step is required. This chapter has introduced the approaches available to pre-concentrate such samples prior to analysis. The techniques described are based on elimination of excess solvent as a means of pre-concentration.

References

1. Karasek, F. W., Clement, R. E. and Sweetman, J. A., *Anal. Chem.*, **53**, 1050A–1058A (1981).
2. Gunther, F. A., Blinn, R. C., Kolbezen, M. J. and Barkley, J. H., *Anal. Chem.*, **23**, 1835–1842 (1951).
3. Ibrahim, E. A., Suffet, I. H. and Sakla, A. B., *Anal. Chem.*, **59**, 2091–2098 (1987).

Chapter 11

Instrumental Techniques for Trace Analysis

Learning Objectives

- To be able to identify the correct analytical technique for the type of inorganic or organic pollutant under investigation.
- To understand the principles of gas chromatography (GC).
- To be able to identify the instrumental requirements for GC.
- To understand the principles of high performance liquid chromatography (HPLC).
- To be able to identify the instrumental requirements for HPLC.
- To understand the relevance of infrared (IR) spectroscopy for total petroleum hydrocarbon analysis.
- To understand the principles of atomic spectroscopy.
- To be able to identify the instrumental requirements for flame atomic absorption spectroscopy (FAAS).
- To be able to identify the instrumental requirements for graphite-furnace atomic absorption spectroscopy (GFAAS).
- To be able to identify the instrumental requirements for hydride-generation atomic absorption spectroscopy (HyFAAS).
- To be able to identify the instrumental requirements for flame photometry (FP).
- To be able to identify the instrumental requirements for inductively coupled plasma–atomic emission spectroscopy (ICP–AES).
- To be able to identify the instrumental requirements for inductively coupled plasma–mass spectrometry (ICP–MS).
- To understand the relevance of anodic stripping voltammetry (ASV) in trace metal analysis.

- To understand the relevance of X-ray fluorescence spectroscopy for the analysis of metals in solid samples.
- To understand the relevance of ion chromatography for the analysis of anions in solution.

11.1 Introduction

After sample preparation comes the analysis. Depending on whether you are looking for organic or inorganic pollutants will determine the choice of analytical technique.

SAQ 11.1

Suggest the appropriate analytical technique(s), from the list below, for the analysis of the following organic and inorganic pollutants.

Pollutant	Technique
Polycyclic aromatic hydrocarbons (PAHs)	
Lead	
Chlorinated phenols	
Chlorinated pesticides	
Arsenic	
Chromium(VI)	
Benzene–toluene–ethylbenzene–xylene(s) (BTEX)	
Total petroleum hydrocarbons (TPHs)	
Nitrate	
Sodium	

Analytical techniques

- gas chromatography with flame ionization detection (GC–FID)
- gas chromatography with electron-capture detection (GC–ECD)
- gas chromatography with mass-selective detection (GC–MSD)
- high performance liquid chromatography with ultraviolet/visible detection (HPLC–UV/vis)
- high performance liquid chromatography with fluorescence detection (HPLC–FL)
- Fourier-transform infrared (FTIR) spectroscopy
- flame atomic absorption spectroscopy (FAAS)
- graphite-furnace atomic absorption spectroscopy (GFAAS)
- hydride-generation atomic absorption spectroscopy (HyFAAS)
- flame photometry (FP)

- inductively coupled plasma–atomic emission spectroscopy (ICP–AES)
- inductively coupled plasma–mass spectrometry (ICP–MS)
- X-ray fluorescence (XRF) spectroscopy
- anodic stripping voltammetry (ASV)
- ion chromatography (IC) with conductivity detection

11.2 Environmental Organic Analysis

11.2.1 Chromatographic Techniques

Environmental organic compounds can be analysed by a variety of techniques, including chromatographic and spectroscopic methods. However, in this present section the main focus will be on the use of chromatographic approaches. It is also not the intention here to give a complete and detailed study of chromatographic analysis but to provide a general overview of the types of separation frequently used in environmental organic analysis. The two most common approaches for separation of an analyte from other compounds in the sample extract are *gas chromatography* (GC) and *high performance liquid chromatography* (HPLC).

DQ 11.1
What is the essential difference between GC and HPLC?

Answer

The essential difference between the two techniques is the nature of the partitioning process. In GC, the analyte is partitioned between a stationary phase and a gaseous phase, whereas in HPLC the partitioning process occurs between a stationary phase and a liquid phase. Separation is therefore achieved in both cases by the affinity of the analyte of interest with the stationary phase. The higher the affinity, then the more the analyte is retained by the column.

The choice of which technique is employed is largely dependent upon the analyte of interest. For example, if the analyte is thermally labile, does not volatilize at temperatures up to 250°C and is strongly polar, then GC is not the appropriate technique. However, HPLC can then be used (and vice versa).

11.2.1.1 Gas Chromatography
Separation in GC is based on the vapour pressures of volatilized compounds and their affinities for the liquid stationary phase, which coats a solid support, as they pass down the column in a carrier gas. The practice of GC can be divided into two broad categories, i.e. packed- and capillary-column based. For

the purpose of this discussion, only capillary-column GC will be discussed. A gas chromatograph (Figure 11.1) consists of a column, typically 15–30 m long, with an internal diameter of 0.1–0.3 mm. The range of column types available from manufacturers is considerable; however, some common types are frequently encountered e.g. the DB-5 type (Figure 11.2).

DQ 11.2

Why do you think that this column is known as a DB-5?

Answer

The DB-5 is a low-polarity column in which the stationary phase, consisting of 5% phenyl- and 95% methylsilicones, is chemically bonded onto silica. In a similar manner, a DB-1 column consists of 100% methyl-silicone.

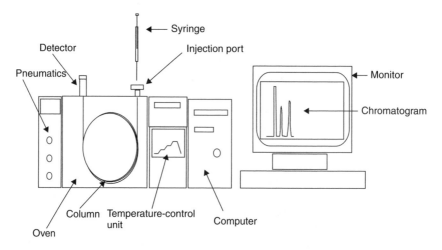

Figure 11.1 Schematic diagram of a typical gas chromatograph. Reproduced by permission of Mr E. Ludkin, Northumbria University, Newcastle, UK.

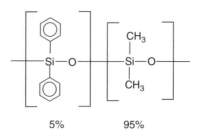

Figure 11.2 The stationary phase of a DB-5 GC column, consisting of 5% diphenyl- and 95% dimethylpolysiloxanes.

The thickness of the stationary phase is of the order of 0.25 µm. The column, through which the carrier gas passes, is placed in a temperature-controlled oven. The carrier gas (e.g. nitrogen) is supplied from a cylinder.

DQ 11.3

As the column is located within an oven, can you suggest two methods of operating the gas chromatograph?

Answer

The gas chromatograph can be operated in either an isothermal or a temperature-programmed mode. In isothermal operation, the oven temperature is constant, e.g. 100°C, throughout the chromatographic run. With temperature programming, the temperature increases from the start to the finish. The temperature-programmed mode is preferred for separation of complex mixtures.

Temperature programming allows a low initial temperature to be maintained to allow the separation of high-boiling-point analytes; this is then followed by a stepwise or linear temperature increase to separate analytes with lower boiling points. Typical column temperature changes can range from 50 to 250°C, at a ramp rate of $7°C\,min^{-1}$. Introduction of the sample requires an injector, of which there are several types. The aim of using an injector is to introduce a small but representative portion of the sample onto the column without overloading. Sample is introduced into the injector by means of a hyperdermic syringe. Two common approaches are applied. The first of these allows a larger sample volume (µl) to be introduced into the injection port and then "splits" or divides the sample – the split/splitless injector (Figure 11.3). In this case, a large volume of sample is introduced into the heated injection port where it is instantly vaporized, with only a small proportion being introduced into the column, and the rest being vented to waste. The ratio of the split flow to the column flow is called the split ratio and can be of the order of 50:1 or 100:1. The other type of injector introduces a smaller volume (nl) of sample directly onto the column – the cold on-column injector (Figure 11.4).

DQ 11.4

What do you think is the major difference in the ways that a split/splitless injector and an on-column injector introduce samples into a gas chromatograph?

Answer

In the case of an on-column injector, a syringe fitted with a very long thin needle is used to introduce all of the sample directly onto the column.

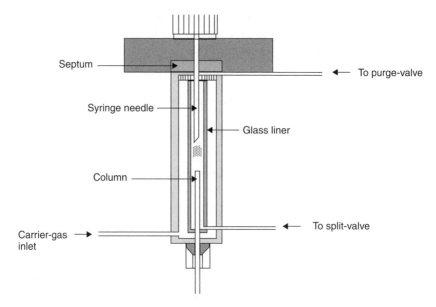

Figure 11.3 Schematic diagram of a split/splitless injector used in gas chromatography. Reproduced by permission of Mr E. Ludkin, Northumbria University, Newcastle, UK.

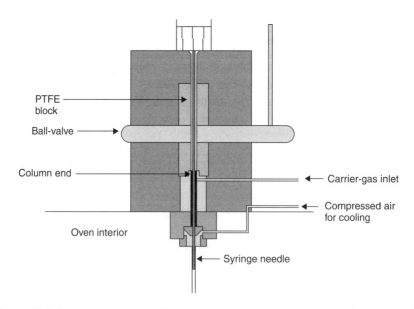

Figure 11.4 Schematic diagram of an on-column injector used in gas chromatography. Reproduced by permission of Mr E. Ludkin, Northumbria University, Newcastle, UK.

The variety of detectors available range from the universal (flame ionization detector) to the specific (electron-capture, thermionic, flame photometric and atomic emission detectors). For example, the electron-capture detector is specific for halogen-containing compounds, e.g. organochlorine pesticides.

DQ 11.5

Can you name an organochlorine pesticide?

Answer

How about one of the following – dieldrin, aldrin or lindane.

In addition, gas chromatography has been coupled to mass spectrometry to provide a highly sensitive detector which also provides information on the molecular structure of the analyte. For a more detailed discussion of both the theoretical aspects and practical details of this approach, the reader is referred to Section 12.2 (Selected Resources) later in this text.

11.2.1.2 High Performance Liquid Chromatography

High performance liquid chromatography can also be classified into two broad categories, i.e. normal phase and reversed phase. For the purpose of this discussion, the most popular of these categories, namely reversed-phase HPLC, will only be discussed. In reversed-phase HPLC, the stationary phase is non-polar, while the mobile phase is polar.

DQ 11.6

What do you think the polarities of the stationary and mobile phases are in normal-phase HPLC?

Answer

In normal-phase HPLC, the stationary phase is polar and the mobile phase is non-polar.

A high performance liquid chromatograph (Figure 11.5) basically consists of a column, typically 25 cm long with an internal diameter of 4.6 mm, packed with a suitable stationary phase (octadecylsilyl (ODS), which consists of a C18 hydrocarbon chain bonded to silica particles of 5–10 μm diameter) through which is passed a mobile phase. The latter is a water–organic solvent system, with typical solvents being methanol or acetonitrile, which is pumped by using a reciprocating or piston pump at a flow rate of 1 ml min^{-1}. The column is normally located in an oven which is maintained at ca. 30°C.

DQ 11.7

Why do you think the column oven is maintained at approximately 30°C?

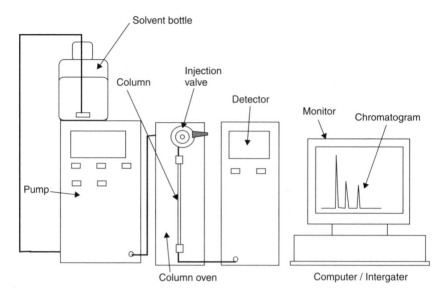

Figure 11.5 Schematic diagram of a typical high performance liquid chromatograph. Reproduced by permission of Mr E. Ludkin, Northumbria University, Newcastle, UK.

Answer

This is to prevent changes in the retention times of the compounds, both between and during chromatographic runs. The temperature of 30°C is arbitrarily fixed to be just above ambient room temperature. It can obviously be adjusted depending on the ambient air temperature.

Samples (10–20 μl) are injected, via a fixed-volume loop connected to a six-port injection valve (Figure 11.6), onto the column and after separation are detected. The most common detector used for HPLC is the ultraviolet–visible spectrometer (available as a single-wavelength unit or with a photodiode array which allows multiple wavelength detection), although a range of more specialized detectors are also available, e.g. fluorescence, electrochemical, refractive index, light-scattering or chemiluminescence. Recently, the introduction of low-cost bench-top liquid chromatograph–mass spectrometers has made the use of this universal detector, with the capability of mass spectral interpretation of unknowns, more readily available.

DQ 11.8

An HPLC system can be operated in two different modes. What are they?

Answer

*The HPLC system can be operated in either the **isocratic mode**, i.e. with the same mobile phase composition throughout the chromatographic run,*

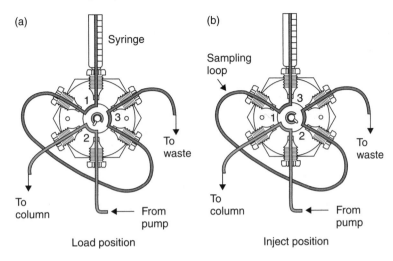

Figure 11.6 Schematic diagrams of a typical injection valve used for high performance liquid chromatography: (a) load position; (b) inject position.

*or by **gradient elution**, i.e. the mobile phase composition varies with respect to the run time. The choice of gradient or isocratic operation depends largely on the number of analytes to be separated and the speed with which the separation is required to be achieved.*

11.2.2 Other Techniques

Infrared (IR) spectroscopy is regularly used for the identification of compounds, often in conjunction with nuclear magnetic resonance (NMR) spectroscopy and mass spectrometry (MS). However, it can also be used for the *quantitative* analysis of environmental compounds, e.g. BTEX, in a sample extract.

DQ 11.9

What is the basis of infrared spectroscopy?

Answer

This technique is concerned with the energy changes involved in the stretching and bending of covalent bonds in molecules.

Infrared spectra are represented in terms of a plot of percentage transmittance versus wavenumber (cm^{-1}). In its most common form, infrared spectroscopy makes use of *Fourier transformation*, a procedure for interconverting frequency functions and time or distance functions. Fourier-transform IR (FTIR) spectroscopy allows the rapid scanning of spectra, with great sensitivity, coupled with simplicity of operation. FTIR spectra can be obtained for solid, liquid or gaseous samples by the use of an appropriate sample cell. Spectra of liquid samples are normally obtained by placing the pure, dry sample between two sodium chloride discs (plates) and placing this in the path of the IR radiation. Solid samples can be prepared as either a Nujol® mull (where a finely ground solid is mixed with a liquid paraffin) and placed between two sodium chloride discs or as a KBr disc (where finely ground powder is mixed with potassium bromide and pressed as a pellet). These sample preparation techniques are all used for qualitative analysis. For *quantitative* work in environmental analysis, typically a sample extract, the liquid extract is placed in a solution cell which is then positioned in the path of the IR radiation. For example, the analysis of BTEX in a suitable extract can be determined by observing the IR spectrum at approximately 3000 cm^{-1} (C–H stretching frequencies occur at $>3000 \text{ cm}^{-1}$ in unsaturated systems while at $< 3000 \text{ cm}^{-1}$ C–H stretching frequencies occur for CH_3, CH_2 and CH groups in saturated systems). By recording the percentage transmittance (signal) for a range of standards and plotting a graph of concentration versus signal, unknown concentrations can then be determined. Care should be taken to ensure that a suitable solvent has been used to extract the sample. After all, the FTIR spectra will display signals for C–H groups in solvents, as well as those of the sample extracts. In this context, a standard EPA method exists for the FTIR analysis of total recoverable petroleum hydrocarbons in environmental samples (EPA Method 8440) after supercritical fluid extraction using CO_2 only (EPA Method 3560). Other suitable solvents include 'Freon-113' and carbon tetrachloride, although the use of these is no longer recommended. The manufacture of 'Freon-113' is no longer possible under the *Montreal Protocol on Substances that Deplete the Ozone Layer* (1990), while carbon tetrachloride is a known carcinogen and also has ozone-depleting properties. For these reasons, the preferred choice, after supercritical CO_2, is tetrachloroethene (or perchloroethylene), C_2Cl_4,

11.3 Environmental Inorganic Analysis

The analysis of inorganic compounds can be carried out by using a variety of analytical techniques, including those based on atomic spectroscopy, X-ray fluorescence spectroscopy, mass spectrometry, electrochemical approaches and chromatography. However, for the purpose of this present section the techniques chosen to be highlighted are those based on atomic spectroscopy, including atomic absorption and atomic emission spectroscopies. Some brief description on the use of an inductively coupled plasma for inorganic mass spectrometry will also be covered.

11.3.1 Atomic Absorption Spectroscopy

Atomic absorption spectroscopy (AAS) is probably the most commonly encountered of the techniques in the laboratory due to the simplicity of the method and

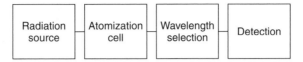

Figure 11.7 Block diagram of a typical atomic absorption spectrometer. From Dean, J. R., *Atomic Absorption and Plasma Spectroscopy*, ACOL Series, 2nd Edn, Wiley, Chichester, UK, 1997. Reproduced with permission of the University of Greenwich.

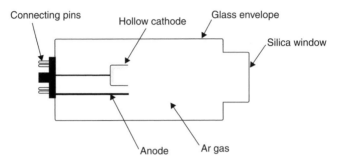

Figure 11.8 Schematic diagram of a hollow-cathode lamp used in atomic absorption spectroscopy. From Dean, J. R., *Atomic Absorption and Plasma Spectroscopy*, ACOL Series, 2nd Edn, Wiley, Chichester, UK, 1997. Reproduced with permission of the University of Greenwich.

its low capital cost. The main components of an atomic absorption spectrometer are a radiation source, an atomization cell and a method of wavelength selection and detection (Figure 11.7). The radiation source, or hollow-cathode lamp (HCL) (Figure 11.8), generates a characteristic narrow-line emission of a selected metal.

DQ 11.10

If you were analysing for lead in a sample, which hollow-cathode lamp would you be using?

Answer

You would, of course, use a lead hollow-cathode lamp.

The atomization cell is the site were the sample is introduced; the type of atomization cell can vary (flame or graphite-furnace) but essentially it causes the metal-containing sample to be dissociated, such that metal atoms are liberated from a hot environment. Such an environment of the atomization cell is sufficient to cause a broadening of the absorption line of the metal. By utilizing the narrowness of the emission line from the radiation source, together with the broad absorption line, means that the wavelength selector only has to isolate the line of interest from other lines emitted by the radiation source (Figure 11.9). This

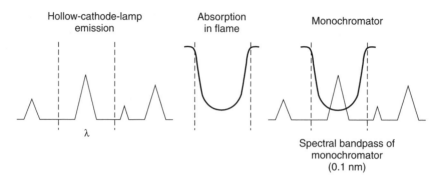

Hollow-cathode-lamp
emission

Absorption
in flame

Monochromator

λ

Spectral bandpass of
monochromator
(0.1 nm)

Figure 11.9 Basic principle of atomic absorption spectroscopy – the 'lock and key' effect. From Dean, J. R., *Atomic Absorption and Plasma Spectroscopy*, ACOL Series, 2nd Edn, Wiley, Chichester, UK, 1997. Reproduced with permission of the University of Greenwich.

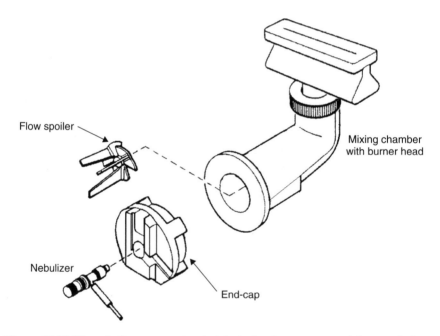

Flow spoiler

Mixing chamber
with burner head

Nebulizer

End-cap

Figure 11.10 The nebulizer–expansion (mixing) chamber system used for sample introduction in flame atomic absorption spectroscopy.

unique feature of AAS gives it a high degree of selectivity; this process is usually referred to as the 'lock and key' effect.

The most common atomization cell is the pre-mixed laminar flame. In this case, the fuel and oxidant gases are mixed prior to entering the burner (the ignition site) in an expansion chamber (Figure 11.10). Two flames are usually

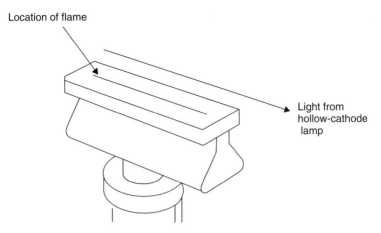

Figure 11.11 Schematic diagram of the slot-burner used in flame atomic absorption spectroscopy.

used in AAS, i.e. either the air–acetylene flame or the nitrous oxide–acetylene flame. Both are located in a slot burner which is positioned in the lightpath of the HCL (Figure 11.11). The choice of flame is straightforward – the air–acetylene flame (slot length, 100 mm) is the most commonly used, whereas the nitrous oxide–acetylene flame (slot length, 50 mm) is reserved for the more refractory elements, e.g. Al. The latter choice for such elements may well indicate the main characteristic difference of the flames i.e. temperature.

DQ 11.11

From the information already given, which do you think is the hotter flame?

Answer

The nitrous oxide–acetylene flame, with the smaller burner slot length, is the hotter flame (3150 K) compared to the air–acetylene flame (2500 K).

The introduction of an aqueous sample into the flame is achieved by using a pneumatic concentric nebulizer/expansion chamber arrangement. The nebulizer (Figure 11.12) consists of a concentric stainless-steel tube through which a Pt/Ir capillary tube is located. The sample is drawn up through the capillary by the action of the oxidant gas (air) escaping through the exit orifice that exists between the outside of the capillary tube and the inside of the stainless-steel concentric tube. The action of the escaping air and liquid sample is sufficient to break it up into a coarse aerosol. This action is called the *Venturi effect*.

DQ 11.12

What do you think are the dual functions of the expansion chamber?

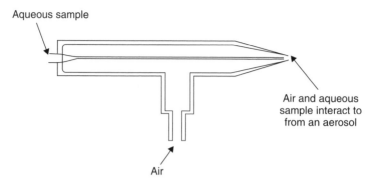

Figure 11.12 Schematic diagram of the pneumatic concentric nebulizer used in atomic absorption spectroscopy. From Dean, J. R., *Atomic Absorption and Plasma Spectroscopy*, ACOL Series, 2nd Edn, Wiley, Chichester, UK, 1997. Reproduced with permission of the University of Greenwich.

Answer

The first is to convert the aqueous sample solution into a coarse aerosol using the oxidant gas, and to then allow this aerosol to be dispersed into a finer aerosol for transport to the burner for atomization or allow residual aerosol particles to condense and go to waste. Secondly, the arrangement also allows safe pre-mixing of the oxidant and fuel gases in the expansion chamber, prior to introduction into the laminar flow burner.

Another type of atomization cell is the graphite furnace. This is used when only a small amount of sample is available or when an increase in sensitivity is required. The graphite atomizer replaces the flame/burner arrangement in the atomic absorption spectrometer. The principal of operation is that a small discrete sample (5–100 μl) is introduced onto the inner surface of a graphite tube through a small opening (Figure 11.13). The graphite tube is arranged so that light from the HCL passes directly through it. This tube is 3–5 cm long with a diameter of 3–8 mm. Heating of the graphite tube is achieved by the passage of an electric current through the tube via water-cooled contacts at each end. Careful control of the heating allows various stages to be incorporated into the programmable heating cycle. Various stages of heating (Figure 11.14) are required to dry the sample, remove the sample matrix and finally to atomize the analyte. An additional heating cycle may be introduced for cleaning of residual material. It is the manner of these heating cycles that is the key to the success of this technique.

DQ 11.13

See if you can elaborate on the different functions of the heating stages and suggest appropriate temperatures when they might occur.

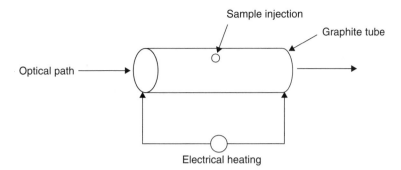

Figure 11.13 Schematic diagram of a typical graphite furnace used in atomic absorption spectroscopy. From Dean, J. R., *Atomic Absorption and Plasma Spectroscopy*, ACOL Series, 2nd Edn, Wiley, Chichester, UK, 1997. Reproduced with permission of the University of Greenwich.

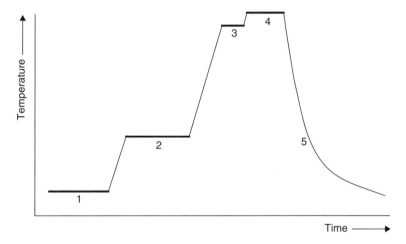

Figure 11.14 A typical time–temperature profile for graphite-furnace atomic absorption spectroscopy: 1, drying; 2, ashing; 3, atomization; 4, cleaning; 5, cooling. From Dean, J. R., *Atomic Absorption and Plasma Spectroscopy*, ACOL Series, 2nd Edn, Wiley, Chichester, UK, 1997. Reproduced with permission of the University of Greenwich.

Answer

The drying stage is necessary to remove residual solvent from the sample. This can be achieved by maintaining the heat of the graphite tube above the boiling point of the solvent, e.g. for water at 110°C for 30 s. The second step involves destruction of the sample matrix in a process called 'ashing'. This involves heating the tube between 350 and 1200°C for ca. 45 s. In an ideal situation, organic matrix components are removed *without any loss of the analyte of interest. As can be appreciated, this is not always possible and this is one of the major disadvantages of the technique. Finally, the temperature of the graphite tube is raised to between 2000 and 3000°C for 2–3 s, to allow atomization of the analyte of interest. It is only during this final atomization step that the absorption of the radiation source by the atomic vapour is measured. It is common to have an internal gas flow of an inert gas (N_2 or Ar) during the drying and ashing stages to remove the extraneous material.*

A specialized form of atomization cell is available for a limited number of elements that are capable of forming volatile hydrides (e.g. As, Bi, Sb, Se and Sn). In this situation, an acidified sample solution is reacted with a sodium tetraborohydride solution. After a short time, the gaseous hydride is liberated.

Various gas–liquid separation devices have been used for either batch or on-line separation. Whichever separation device is used, the improvement in sensitivity is high. The generated hydride is transported to the atomization cell by using a carrier gas. This cell consists of either an electrically heated or flame-heated quartz tube for atomization. In addition, cold-vapour generation is exclusively reserved for the element mercury. In this situation, the mercury present in the sample is reduced, usually using tin(II) chloride, to elemental mercury. The mercury vapour generated is then transported to the atomization cell by a carrier gas. The cell is in the form of a long-pathlength glass absorption cell located in the path of the HCL. Mercury is monitored at 253.7 nm.

Wavelength separation is achieved by using a monochromator. The Czerny–Turner monochromator has a focal length of 0.25–0.5 m with a grating containing only 600 lines mm^{-1} and a resolution of 0.2–0.02 nm. The attenuation of the HCL radiation by the atomic vapour is detected by a photomultiplier tube (PMT). The latter is a device for converting, proportionally, photons to current. Incident light strikes a photosensitive material which converts the light into electrons (the photoelectric effect). The generated electrons are then focused and multiplied by a series of dynodes prior to collection at the anode. The multiplied electrons (or electrical current) are converted into a voltage signal which is then sent via an analogue-to-digital (A/D) converter to a suitable computer for processing.

The occurrence of molecular absorbance and scatter in AAS can be overcome by the use of background correction methods. Various types of correction procedures are common, e.g. continuum source, Smith–Hieftje and the Zeeman effect. In addition, other problems can occur and include those based on chemical, ionization, physical and spectral interferences.

11.3.2 Atomic Emission Spectroscopy

The instrumentation used for atomic emission spectroscopy (AES) consists of an atomization cell, a spectrometer/detector and a read-out device. In its simplest form, flame photometry (FP), the atomization cell consists of a flame (e.g.

air–natural gas), while the spectrometer comprises an interference filter followed by a photodiode or photoemissive detector. Flame photometry is used for the determination of, for example, potassium (766.5 nm) or sodium (589.0 nm).

Most modern instruments for AES use an inductively coupled plasma (ICP) as the atomization cell. The ICP is formed within the confines of three concentric glass tubes or plasma torch (Figure 11.15). Each concentric glass tube has an entry point, with the intermediate (plasma) and external (coolant) tubes having tangentially arranged entry points and the inner tube consisting of a capillary tube through which the aerosol is introduced from the nebulization/spray chamber. Located around the outer glass tube is a coil of copper tubing through which water is recirculated. Power input to the ICP is achieved through this copper load or induction coil, typically in the range 0.5–1.5 kW at a frequency of 27 or 40 MHz. The inputted power causes the induction of an oscillating magnetic field whose lines of force are axially orientated inside the plasma torch and follow

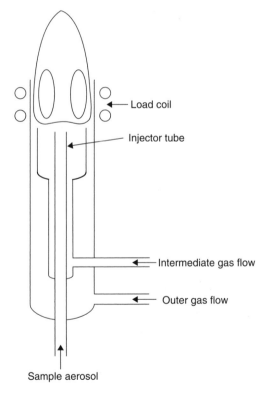

Figure 11.15 Schematic diagram of an inductively coupled plasma located within its torch, as employed in atomic emission spectroscopy. From Dean, J. R., *Atomic Absorption and Plasma Spectroscopy*, ACOL Series, 2nd Edn, Wiley, Chichester, UK, 1997. Reproduced with permission of the University of Greenwich.

elliptical paths outside the induction coil. At this point in time, no plasma exists. In order to initiate the plasma, the carrier gas flow is first switched off and a spark added momentarily from a Tesla coil, which is attached to the outside of the plasma torch by means of a piece of copper wire. Instantaneously, the spark, a source of 'seed' electrons, causes ionization of the argon carrier gas. This process is self-sustaining so that argon, argon ions and electrons co-exist within the confines of the plasma torch but protruding from the top in the shape of a bright white luminous 'bullet'. This characteristic bullet shape is formed by the escaping high-velocity argon gas causing air entrainment back towards the plasma torch itself. In order to introduce the sample aerosol into the confines of the hot plasma gas (7000–10 000 K) the carrier gas is switched on; this punches a hole into the centre of the plasma, thus creating the characteristic 'doughnut' or toroidal shape of the ICP. In the conventional ICP system, the emitted radiation is viewed laterally, or side-on. Therefore, the element radiation of interest is 'viewed' through the luminous plasma.

DQ 11.14

Do you think that viewing the plasma side-on will have any detrimental effect?

Answer

As you are viewing the elemental radiation side-on you must also be raising the level of background radiation observed. Some modern instruments allow the plasma to be viewed end-on, thereby reducing the background radiation.

The most common method of liquid sample introduction in AES is via a nebulizer. The type of nebulizer used in modern instruments has not altered significantly since its first usage, despite of its inefficiency. Most nebulizers have transport efficiencies of between 1–2%.

DQ 1.15

How would you define transport efficiency?

Answer

Transport efficiency is defined as the amount of the original sample solution that is converted to an aerosol and then transported into the plasma source.

The basis of the nebulizer is to convert an aqueous sample into an aerosol by the action of a carrier gas. In order to produce an aerosol of sufficient particle size, ideally of <10 μm to avoid substantial cooling/extinguishing of the plasma, it is necessary to present the generated aerosol into a spray chamber. The latter has the advantage of further reducing the original aerosol particle size towards the ideal by providing a surface for collisions and/or condensation. The generation of condensation represents part of the inefficiency of the nebulizer/spray chamber sample-introduction system.

Light emitted from the plasma source is focused onto the entrance slit of a spectrometer by using a convex lens arrangement. The spectrometer is required to separate the emitted light into its component wavelengths. In practice, depending on the requirements of the analyst and the capital cost of the instrument, two options are available. The first involves a capability to measure one wavelength, corresponding to one element at a time, while the second allows multi-wavelength or multi-element detection. The former is called *sequential analysis* or *sequential multi-element analysis* if the system is to be used to measure several wavelengths one at a time, while the latter is termed *simultaneous multi-element analysis*. The typical wavelength coverage of a spectrometer for AES is between 167 nm (Al) to 852 nm (Cs). After wavelength separation has been achieved, it is obviously necessary to 'view' the spectral information. This is most commonly achieved by using either a photomultiplier tube (PMT) or a charge-coupled device (CCD).

Like all spectroscopic techniques, AES suffers from some interferences, e.g. spectral interferences. Such interferences for AES can be classified into two main categories, i.e. spectral overlap and matrix effects. Spectral interferences are probably the most well known and best understood. The usual remedy to alleviate a spectral interference is to either increase the resolution of the spectrometer or to select an alternative spectral emission line.

11.3.3 Inductively Coupled Plasma–Mass Spectrometry

Inductively coupled plasma–mass spectrometry combines the benefits of the ICP with mass spectrometry. The major instrumental development required in order to establish ICP–MS was the efficient coupling of an ICP, operating at atmospheric pressure, with a mass spectrometer, which operates under high vacuum (Figure 11.16). The development of a suitable interface held the key to the establishment of the technique. The only instrumental alteration is in how the analyte is observed. In AES, the ICP torch is positioned vertically, so that emission can (normally) be observed at right angles by the (atomic emission) spectrometer. In MS, the ICP torch is positioned horizontally, so that ions can be extracted from the ICP directly into the mass spectrometer (Figure 11.17). As a consequence of this horizontal positioning of the ICP torch in relation to the mass spectrometer, all species that enter the plasma are transferred into the MS unit.

DQ 11.16

How would you introduce a sample into an ICP–MS system?

Answer

The sample introduction devices used for ICP–AES can be similarly used for ICP–MS.

Mass spectrometer

Plasma

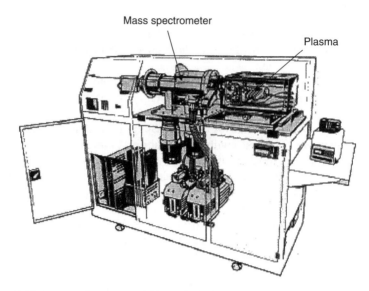

Figure 11.16 Layout of a commercially available inductively coupled plasma–mass spectrometry (ICP–MS) system. From applications literature published by VG Elemental. Reproduced by permission of Thermo Elemental, Winsford, Cheshire, UK.

Slide-valve

Sampling cone

Skimmer cone

ICP

Pressure
~0.0001 mbar   ~2.5 mbar

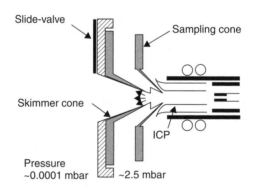

Figure 11.17 Schematic diagram of the inductively coupled plasma/mass spectrometer interface. From Dean, J. R., *Atomic Absorption and Plasma Spectroscopy*, ACOL Series, 2nd Edn, Wiley, Chichester, UK, 1997. Reproduced with permission of the University of Greenwich.

The quadrupole mass spectrometer acts as a filter, transmitting ions with a pre-selected mass/charge ratio. The transmitted ions are then detected with a channel electron multiplier. ICP–MS can be operated in two distinctly different modes, i.e. with the mass filter transmitting only one mass/charge ratio, or where the DC and RF values are changed continuously. The former would allow single-ion monitoring (SIM), while the latter allows multi-element analysis. This permits the analyst to carry out fast repetitive analyses of a pre-determined set of elements.

Quadrupole mass analysers are capable of only unit mass resolution, i.e. they can observe integral values of the mass/charge ratio only (204, 205, 206, etc.). This limited resolution leads to interferences. The type of interferences can be broadly classified, according to their origin, into isobaric, molecular and matrix-dependent. Just as in AES, some overlap of elements can occur (isobaric), and just as in the other case, such information is well documented. However, other types of interference can occur, with these being the result of the acid(s) used to prepare the sample and/or the argon plasma gas (polyatomics). In addition, the formation of oxide, hydroxide and doubly charged species is possible. Finally, the occurrence of matrix-interferences results in signal enhancement or depression with respect to atomic mass.

11.3.4 Other Techniques

Anodic stripping voltammetry (ASV) is an electroanalytical technique used for the analysis of trace metals in solution. The apparatus for this consists of three electrodes located in an electrolytic cell. These electrodes are a working electrode, e.g. a mercury-drop, a reference electrode and a counter electrode. Sample is placed in the cell, together with a supporting electrolyte, e.g. 0.1 mol$\,$l^{-1} acetate buffer at pH 4.5. Dissolved oxygen is removed from the solution by bubbling nitrogen or argon through the cell. By holding the working electrode at a small negative potential (with respect to the reference electrode), the metal ions in solution are attracted to the electrode and deposited. This process is aided by stirring of the solution. By careful control of the deposition time and stirring rate, the amount of metal deposited is proportional to its original concentration. After a specified time, the working electrode potential is slowly changed to become less negative (in the positive direction). At specific potentials, the deposited metal on the surface of the working electrode is oxidized, and hence returned to the solution. This process is monitored by plotting the current change between the working electrode and counter electrode against the potential. The resultant voltammogram (Figure 11.18) can be used to determine the concentration of metals, e.g. lead, copper and zinc, in solution at trace levels.

X-ray fluorescence (XRF) spectroscopy can be used to analyse metals in solids directly. In this technique, sample atoms are irradiated with X-rays, thus causing ejection of electrons from the inner shell. The loss of these lower-energy electrons results in outer-shell electrons filling the vacancies, and hence emitting X-rays characteristic of the sample atom. Two different instruments for XRF are available, dependent upon how the radiation is observed. In wavelength-dispersive XRF, the emitted radiation from the sample is dispersed by a crystal into its component wavelengths. Sequential scanning of the different wavelengths thus allows measurements of the different elements. The alternative approach is to use energy-dispersive XRF. In this situation, all emitted radiation is measured at

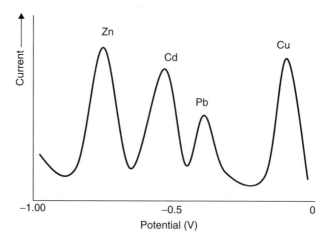

Figure 11.18 Typical example of an anodic stripping voltammogram, obtained for the analysis of trace metals in solution.

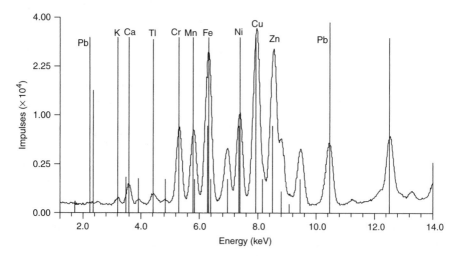

Figure 11.19 Typical example of an X-ray fluorescence trace, obtained for the analysis of trace metals in an environmental solid sample.

the detector and sorted electronically. A typical XRF trace of an environmental sample is shown in Figure 11.19.

Ion chromatography can be used to determine both anions and cations. However, it is its use for anion determination which is of importance in this chapter. Just as in high performance liquid chromatography (described above), separation is achieved by using a column. In this case, the column is likely to be based

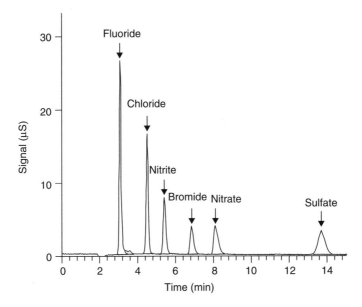

Figure 11.20 Typical ion chromatogram, obtained for the separation/analysis of anions in an environmental sample.

on poly(styrene divinylbenzene), with a mobile phase of sodium hydroxide or a sodium carbonate/hydrogen carbonate buffer. Detection is with a conductivity detector, which measures the increase in conductivity of the eluent containing the sample anions. A typical chromatogram for the separation/determination of anions in a sample is shown in Figure 11.20.

Summary

The final stage in the analytical process is to measure the concentration of the environmental pollutant. This chapter has described appropriate techniques for the measurement of both metals and organic compounds. While the primary descriptions have focused on atomic spectroscopy for metals and chromatography for organic compounds, some related techniques have been discussed briefly.

Chapter 12
Recording of Information in the Laboratory and Selected Resources

Learning Objectives

- To enable relevant scientific information to be recorded in the laboratory.
- To allow correct sample pre-treatment details to be noted.
- To allow correct sample preparation details to be noted.
- To allow correct analytical technique information to be noted.
- To be aware of relevant additional resources that can be consulted, including books, journals, CD-ROMs, videos and the Internet.

12.1 Recording of Information

12.1.1 Introduction

Prior to commencing any experimentation, you are required to complete a hazard and risk assessment of the chemicals and apparatus that you will use, i.e. a Control of Substances Hazardous to Health (COSHH) assessment (see Chapter 1).

When carrying out any laboratory work, it is important to record information in a systematic manner, relating to sample information, sample treatment and the analytical technique being used, calibration strategy and the recording of results. Below are some examples of data sheets that could be used to ensure that all information is recorded in a systematic way. These data sheets are not intended to be totally comprehensive, and so may be altered and amended by the individual worker as required.

The first data sheet (A) allows identification of any preliminary sample pre-treatment that may be necessary prior to sample preparation. Data sheets (B) and (C) identify the sample preparation techniques for inorganic and organic samples, respectively, while data sheets (D) and (E) contain, respectively, the instrumental requirements for the most commonly used inorganic and organic analytical techniques.

12.1.2 Examples of Data Sheets

Data Sheet A: Sample Treatment

- Grinding and sieving
  Grinder used (model/type)...
  Particle size (sieve mesh size)...

- Mixing of the sample
  Manual shaking   yes/no
  Mechanical shaking   yes/no
  Other (specify)...
- Sample storage
  Fridge   yes/no
  Other (specify)...
- Chemical pre-treatment
  pH adjustment   yes/no
  Addition of alkali (specify).............. or acid (specify)..................
  or buffer............................... pH = ...

Data Sheet B: Sample Preparation for Inorganic Analysis

- Sample weight(s)
  (record to four decimal places)

    Sample 1.................. g   Sample 2.................... g

- Acid digestion
  Vessel used...
  Hot-plate or other (specify) ....
  Temperature controlled or not (specify) ....
  Type of acid used...
  Volume of acid used...ml

    Any other details...

- Other method of sample decomposition, e.g. fusion or dry ashing (specify)
  ....
  ....
  ....

- Sample derivatization
  Specify...
  ....
  Reagent concentration.. $mol\,l^{-1}$
  Reagent volume used... ml
  Heat required (specify)..

- Sample dilution
  Specify the dilution factor involved, with appropriate units
  ....
  ....
  ....

- Addition of an internal standard
  Specify...
  Added before digestion   yes/no
  Added after digestion   yes/no

- Sample and reagent blanks
  Specify...
  ....

- Recovery
  Specify...
  Added before digestion   yes/no
  Added after digestion   yes/no

Data Sheet C: Sample Preparation for Organic Analysis

- Sample weight(s)
  (record to four decimal places)

　　Sample 1 .... g　　　　Sample 2 .... g

- Soxhlet extraction method
  Drying agent added and weight (specify) ....
  Type of solvent(s) used ....
  Volume of solvent(s) used .... ml
  Any other details ....

- Other method of sample extraction, e.g. SFE, ASE (PFE) or MAE, plus
  operating conditions (specify)
  ....
  ....
  ....

- Sample clean-up　　　　　　　　　　　　　　　　　　　yes/no
  Specify..
  ....

- Pre-concentration of the sample　　　　　　　　　　　　　yes/no
  Method of solvent reduction...
  Final volume of extract...

- Sample derivatization
  Specify..
  ....
  Reagent concentration... $mol\,l^{-1}$
  Reagent volume used.. ml
  Heat required (specify)..

- Sample dilution
  Specify the dilution factor involved, with appropriate units
  ....
  ....
  ....

- Addition of an internal standard
  Specify..
  Added before extraction　　　　　　　　　　　　　　　　yes/no
  Added after extraction　　　　　　　　　　　　　　　　　yes/no

- Sample and reagent blanks
  Specify..
  ....

- Recovery
  Specify..
  Added before extraction　　　　　　　　　　　　　　　　yes/no
  Added after extraction　　　　　　　　　　　　　　　　　yes/no

Data Sheet D: Analytical Techniques for Inorganic Analysis

Method

ICP–AES   yes/no
ICP–MS   yes/no
FAAS   yes/no
GFAAS   yes/no
Flame photometry   yes/no
XRF   yes/no
Other (specify) ....

Inductively Coupled Plasma–Atomic Emission Spectroscopy

- ICP characteristics
    Manufacturer...
    Frequency... Hz
    Power.. kW
    Observation height................................ mm above load coil

- Argon gas flow rates
    Outer gas flow rate.. $l\,min^{-1}$
    Intermediate gas flow rate....................................... $l\,min^{-1}$
    Injector gas flow rate.. $l\,min^{-1}$

- Sample introduction method
    Nebulizer/spray chamber (specify).......................................
    ....
    ....

- Spectrometer
    Simultaneous or sequential
    Element(s) and wavelength(s)..
    ....

- Quantitation
    Peak height   yes/no
    Peak area   yes/no
    Method used   manual/electronic
    Internal standard...
    External standard..
    Calibration method   direct/standard additions
    Number of calibration standards..
    Linear range of calibration...

Inductively Coupled Plasma–Mass Spectrometry

- ICP characteristics
  Manufacturer...
  Frequency... Hz
  Power... kW

- Argon gas flow rates
  Outer gas flow rate.. $l\,min^{-1}$
  Intermediate gas flow rate... $l\,min^{-1}$
  Injector gas flow rate.. $l\,min^{-1}$

- Sample introduction method
  Nebulizer/spray chamber (specify)....................................
  ....
  ....

- Mass spectrometer
  Element(s) and mass/charge ratio(s)..................................
  ....
  Scanning or peak hopping mode

- Quantitation
  Peak height   yes/no
  Peak area   yes/no
  Method used   manual/electronic
  Internal standard...
  External standard...
  Calibration method   direct/standard additions
  Number of calibration standards.....................................
  Linear range of calibration...

Flame Atomic Absorption Spectroscopy

- Atomizer characteristics
    - Flame...................... Graphite furnace............................
    - Cold-vapour............... Hydride generation...........................

- Flame – gas mixture...
    - Gas flow rates... $l\ min^{-1}$

- Sample introduction method
    - Nebulizer/expansion chamber (specify).....................................
    - ....
    - ....

- Spectrometer
    - Element(s) and wavelength(s)...
    - ....

- Quantitation
    - Peak height   yes/no
    - Peak area   yes/no
    - Method used   manual/electronic
    - Calibration method   direct/standard additions
    - Number of calibration standards...
    - Linear range of calibration...

Electrothermal (Graphite-Furnace) Atomic Absorption Spectroscopy

- Atomizer characteristics
  graphite furnace...

- Furnace characteristics
  Graphite.................... Pyrolytic coated......................
  L'Vov platform..............Other (specify)......................

- Programme
  Drying....................temperature   °C;   time   s
  Ashing....................temperature   °C;   time   s
  Atomization...............temperature   °C;   time   s
  Other.....................temperature   °C;   time   s

- Background correction   yes/no
  Deuterium   yes/no
  Zeeman   yes/no
  Smith–Hieftje   yes/no

- Spectrometer
  Element(s) and wavelength(s)...
  ....

- Quantitation
  Peak height   yes/no
  Peak area   yes/no
  Method used   manual/electronic
  Calibration method   direct/standard additions
  Number of calibration standards...
  Linear range of calibration..

Flame Atomic Emission Spectroscopy (Flame Photometry)

- Atomizer characteristics
  Flame...

- Flame – gas mixture..
  Gas flow rates..

- Sample introduction method
  Nebulizer/expansion chamber (specify)...................................
  ....
  ....

- Spectrometer
  Element(s)..

- Quantitation
  Peak height   yes/no
  Method used   manual/electronic
  Calibration method   direct/standard additions
  Number of calibration standards..
  Linear range of calibration...

X-Ray Fluorescence Spectroscopy
- Spectrometer
  Manufacturer..
  Energy/wavelength-dispersive...

- Sample preparation
  Specify...
  ....
  ....

- Quantitation
  Peak height   yes/no
  Peak area   yes/no
  Internal standard..
  External standard..
  Calibration method   direct/standard additions
  Number of calibration standards..
  Linear range of calibration...

Data Sheet E: Analytical Techniques for Organic Analysis

Method
 GC   yes/no
 HPLC   yes/no
 IC   yes/no
 Other (specify)...

Gas Chromatography

- Column characteristics
 Manufacturer...
 Type...
 Length.................. m; internal diameter....................... mm
 Film thickness.. μm

- Carrier gas and flow rate..

- Isothermal or temperature-programmed   yes/no
 Temperature programme (specify)......................................
 ....
 ....

- Injector type..
 Injector temperature...°C
 Split ratio..
 Injection volume.. μl

- Detector type ....
 Operating conditions..

- Quantitation
 Peak height   yes/no
 Peak area   yes/no
 Method used   manual/electronic
 Internal standard...
 External standard...
 Calibration method   direct/standard additions
 Number of calibration standards..
 Linear range of calibration...

High Performance Liquid Chromatography

- Column characteristics
  Manufacturer...
  Type..
  Reversed-phase or normal-phase
  Length................. cm; internal diameter................... mm
- Mobile phase (solvent)...
  Flow rate..ml min^{-1}

- Isocratic or gradient   yes/no
  Gradient programme (specify)..
  ....
  ....

- Injector volume .... μl

- Detector type...
  Operating conditions...

- Quantitation
  Peak height   yes/no
  Peak area   yes/no
  Method used   manual/electronic
  Internal standard..
  External standard..
  Calibration method: direct/standard additions
  Number of calibration standards:.......................................
  Linear range of calibration..

12.2 Selected Resources

To assist readers to maximize their time, and hence efficiency, a selected resource list is provided.

12.2.1 Journals

Regular articles on the latest advances appear in the following journals:

- The Analyst
- Analytica Chimica Acta
- Analytical Chemistry
- Applied Spectroscopy
- Journal of Analytical Atomic Spectrometry
- Journal of Chromatography, Part A
- Journal of Environmental Monitoring
- Mikrochimica Acta
- Spectrochimica Acta, Part B
- Talanta
- Trends in Analytical Chemistry
- Fresenius' Journal for Analytical Chemistry

12.2.2 Books†

12.2.2.1 Specific Books on Laboratory Safety

Anon, *Safe Practices in Chemical Laboratories*, The Royal Society of Chemistry, London, 1989.
Anon, *COSHH in Laboratories*, The Royal Society of Chemistry, London, 1989.
Lenga, R. E. (Ed.), *Sigma-Aldrich Library of Chemical Safety Data*, 2nd Edn, Sigma–Aldrich Ltd, Gillingam, UK, 1988.

12.2.2.2 Specific Books on General Analytical Chemistry

Currell, G., *Analytical Instrumentation: Performance Characteristics and Quality*, AnTS Series, Wiley, Chichester, UK, 2000.
Mendham, J., Denney R. C., Barnes, J. D. and Thomas M. J. K., *Vogel's Textbook of Quantitative Chemical Analysis*, 6th Edn, Prentice Hall, Harlow, UK, 2000.
Perez-Bendito, D. and Rubio, S., *Environmental Analytical Chemistry*, Elsevier, Amsterdam, 2000.

---

† Arranged in chronological order.

Rouessac, F. and Rouessac, A., *Chemical Analysis: Modern Instrumentation Methods and Techniques*, Wiley, Chichester, UK, 2000.

Rubinson, K. A. and Rubinson, J. F., *Contemporary Instrumental Analysis*, Prentice Hall, Harlow, UK, 2000.

Crawford K. and Heaton A., *Problem Solving in Analytical Chemistry*, The Royal Society of Chemistry, Cambridge, UK, 1999.

Kellner, R., *Analytical Chemistry*, Wiley, Chichester, UK, 1998.

Schwedt, G., *The Essential Guide to Analytical Chemistry*, Wiley, Chichester, UK, 1997.

Caroli, S., *Element Speciation in Bioinorganic Chemistry*, Wiley, Chichester, UK, 1996.

Markert, B. and Zittau, I. H. L., *Instrumental Elemental and Multielement Analysis of Plant Samples: Methods and Applications*, Wiley, Chichester, UK, 1996.

Skoog, D. A., West, D. M. and Holler, F. J., *Fundamentals of Analytical Chemistry*, 7th Edn, Saunders College Publishing, Orlando, FL, 1996.

Harris, D. C., *Quantitative Chemical Analysis*, 4th Edn, Freeman and Co., Inc., New York, 1995.

Prichard, E., *Quality in the Analytical Chemistry Laboratory*, ACOL Series, Wiley, Chichester, UK, 1995.

Seivers, R. E., *Selective Detectors: Environmental, Industrial and Biomedical Applications*, Wiley, Chichester, UK, 1995.

Christian, G., *Analytical Chemistry*, 5th Edn, Wiley, Chichester, UK, 1994.

Parkany, M. (Ed.), *Quality Assurance for Analytical Laboratories*, The Royal Society of Chemistry, Cambridge, UK, 1994.

Crompton, T. R., *The Analysis of Natural Waters*, Vol. 1, *Complex-Formation Preconcentration Techniques*, Oxford University Press, Oxford, UK, 1993.

Crompton, T. R., *The Analysis of Natural Waters*, Vol. 2, *Direct Preconcentration Techniques*, Oxford University Press, Oxford, UK, 1993.

Kateman, G. and Buydens, L., *Quality Control in Analytical Chemistry*, Wiley, Chichester, UK, 1993.

Skoog, D. A. and Leary, J. L., *Principles of Instrumental Analysis*, 4th Edn, Saunders College Publishing, Orlando, FL, 1992.

Crompton, T. R., *Analytical Instrumentation for the Water Industry*, Butterworth-Heinemann Ltd, Oxford, UK, 1991.

Day R. A. and Underwood, A. L., *Quantitative Analysis*, 6th Edn, Prentice Hall, Harlow, UK, 1991.

Broekaert, J. A. C., Gucer, S. and Adams, F., *Metal Speciation in the Environment*, NATO ASI Series, Vol. G23, Springer-Verlag, Berlin, 1990.

Busch, K. W. and Busch, M. A., *Multielement Detection Systems for Spectrochemical Analysis*, Wiley, New York, 1990.

Willard, H. H., Meritt, Jr, L. L., Dean, J. A. and Settle, Jr, F. A., *Instrumental Methods of Analysis*, 7th Edn, Wadsworth Publishing Co., Belmont, CA, 1988.

12.2.2.3 Specific Books on Atomic Spectroscopy

Beauchemin, D., Gregoire, D. C., Karanassios, V., Wood, T. J. and Mermet, J. M., *Discrete Sample Introduction Techniques for Inductively Coupled Plasma–Mass Spectrometry*, Elsevier, Amsterdam, 2000.

Caruso, J. A., Sutton, K. L. and Ackley, K. L., *Elemental Speciation: New Approaches for Trace Element Analysis*, Elsevier, Amsterdam, 2000.

Taylor, H., *Inductively Coupled Plasma–Mass Spectrometry: Practices and Techniques*, Academic Press, London, 2000.

Ebdon, L., Evans, H., Fisher, A. and Hill, S., *An Introduction to Atomic Absorption Spectrometry*, Wiley, Chichester, UK, 1998.

Dean, J. R., *Atomic Absorption and Plasma Spectroscopy*, 2nd Edn, ACOL Series, Wiley, Chichester, UK, 1997.

Lobinski, R. and Marczenko, Z., *Spectrochemical Trace Analysis for Metals and Metalloids*, in *Wilson and Wilsons Comprehensive Analytical Chemistry*, Vol. XXX, Weber, S. G. (Ed.), Elsevier, Amsterdam, 1996.

Cresser, M. S., *Flame Spectrometry in Environmental Chemical Analysis: A Practical Approach*, The Royal Society of Chemistry, Cambridge, UK, 1995.

Dedina, J. and Tsalev, D. I., *Hydride Generation Atomic Absorption Spectrometry*, Wiley, Chichester, UK, 1995.

Evans, E. H., Giglio, J. J., Castillano, T. M. and Caruso, J. A., *Inductively Coupled and Microwave Induced Plasma Sources for Mass Spectrometry*, The Royal Society of Chemistry, Cambridge, UK, 1995.

Fang, Z., *Flow Injection Atomic Absorption Spectrometry*, Wiley, Chichester, UK, 1995.

Howard, A. G. and Statham, P. J., *Inorganic Trace Analysis: Philosophy and Practice*, Wiley, Chichester, UK, 1993.

Slickers, K., *Automatic Atomic Emission Spectroscopy*, 2nd Edn, Bruhlsche Universitatsdruckerei, Giessen, Germany, 1993.

Vandecasteele. C. and Block, C. B., *Modern Methods of Trace Element Determination*, Wiley, Chichester, UK, 1993.

Jarvis, K. E., Gray, A. L. and Houk, R. S., *Handbook of Inductively Coupled Plasma Mass Spectrometry*, Blackie and Son, Glasgow, UK, 1992.

Lajunen, L. H. J., *Spectrochemical Analysis by Atomic Absorption and Emission*, The Royal Society of Chemistry, Cambridge, UK, 1992.

Holland, G. and Eaton, A. N., *Applications of Plasma Source Mass Spectrometry*, The Royal Society of Chemistry, Cambridge, UK, 1991.

Jarvis, K. E., Gray, A. L., Jarvis, I. and Williams, J. G., *Plasma Source Mass Spectrometry*, The Royal Society of Chemistry, Cambridge, UK, 1990.

Sneddon, J., *Sample Introduction in Atomic Spectroscopy*, Vol. 4, Elsevier, Amsterdam, 1990.

Date, A. R. and Gray, A. L., *Applications of Inductively Coupled Plasma Mass Spectrometry*, Blackie and Son, Glasgow, UK, 1989.

Harrison, R. M. and Rapsomanikis, S., *Environmental Analysis Using Chromatography Interfaced with Atomic Spectroscopy*, Ellis Horwood Ltd, Chichester, UK, 1989.

Moenke-Blankenburg, L., *Laser Microanalysis*, Wiley, New York, 1989.

Moore, G. L., *Introduction to Inductively Coupled Plasma Atomic Emission Spectroscopy*, Elsevier, Amsterdam, 1989.

Thompson, M. and Walsh, J. N., *A Handbook of Inductively Coupled Plasma Spectrometry*, 2nd Edn, Blackie and Son, Glasgow, UK, 1989.

Adams, F., Gijbels, R. and van Grieken, R., *Inorganic Mass Spectrometry*, Wiley, New York, 1988.

Ingle, J. D. and Crouch, S. R., *Spectrochemical Analysis*, Prentice-Hall International, London, 1988.

Boumans P. W. J. M., *Inductively Coupled Plasma Emission Spectrometry*, Parts 1 and 2, Wiley, New York, 1987.

Montaser, A. and Golightly, D. W., *Inductively Coupled Plasmas in Analytical Atomic Spectrometry*, VCH, New York, 1987.

Welz, B., *Atomic Absorption Spectrometry*, VCH, Weinheim, Germany, 1985.

Ebdon, L., *An Introduction to Atomic Absorption Spectroscopy*, Heyden and Son, London, 1982.

12.2.2.4 Specific Books on Chromatography

Grob, K., *Split and Splitless Injection in Capillary Gas Chromatography*, 4th Edn, Wiley, Chichester, UK, 2001.

Kleibohmer, W. (Ed.), *Handbook of Analytical Separations*, Vol. 3, *Environmental Analysis*, Elsevier, Amsterdam, 2001.

Ahuja, S., *Handbook of Bioseparations*, Academic Press, London, 2000.

Fritz, J. S. and Gjerde, D. T., *Ion Chromatography*, 3rd Edn, Wiley, Chichester, UK, 2000.

Hahn-Deinstrop, E., *Applied Thin Layer Chromatography*, Wiley, Chichester, UK, 2000.

Hubschmann, H. J., *Handbook of GC/MS*, Wiley, Chichester, UK, 2000.

Kromidas, S., *Practical Problem Solving in HPLC*, Wiley, Chichester, UK, 2000.

Subramanian, G., *Chiral Separation Techniques*, 2nd Edn, Wiley, Chichester, UK, 2000.

Weinberger, R., *Practical Capillary Electrophoresis*, 2nd Edn, Academic Press, London, 2000.

Wilson, I., *Encyclopedia of Separation Science*, Academic Press, London, 2000.

Hanai, T., *HPLC: A Practical Guide*, The Royal Society of Chemistry, Cambridge, UK, 1999.

Jennings, W., Mittlefehld, E. and Stremple, P., *Analytical Gas Chromatography*, 2nd Edn, Academic Press, London, 1997.

Kolb, B. and Ettre, L. S., *Static Headspace–Gas Chromatography: Theory and Practice*, Wiley-Interscience, New York, 1997.

Synder, L. R., Kirkland, J. J. and Glajch J. L., *Practical HPLC Method Development*, 2nd Edn, Wiley, New York, 1997.
Weston, A. and Brown, P., *HPLC and CE: Principles and Practice*, Academic Press, London, 1997.
Fowlis, I. A., *Gas Chromatography*, 2nd Edn, ACOL Series Wiley, Chichester, UK, 1995.
Robards, K., Haddard, P. R. and Jackson, P. E., *Principles and Practice of Modern Chromatographic Methods*, Academic Press, London, 1994.
Braithwaite, A. and Smith, F. J., *Chromatographic Methods*, 4th Edn, Chapman & Hall, London, 1990.
Schoenmakers, P. J., *Optimization of Chromatographic Selectivity: A Guide to Method Development*, Elsevier, Amsterdam, 1986.
Synder, L. R. and Kirkland, J. J., *Introduction to Modern Liquid Chromatography*, 2nd Edn, Wiley, New York, 1979.

12.2.2.5 Specific Books on Electroanalytical Techniques

Monk, P. M. S., *Fundamentals of Electroanalytical Chemistry*, AnTS Series, Wiley, Chichester, UK, 2000.
Wang, J., *Analytical Electrochemistry*, 2nd Edn, Wiley, Chichester, UK, 2000.

12.2.2.6 Specific Books on Sample Preparation

Williams, J. R. and Clifford, A. A., *Supercritical Fluids: Methods and Protocols*, Humana Press, Totowa, NJ, 2000.
Handley, A. J., *Extraction Methods in Organic Analysis*, Sheffield Academic Press, Sheffield, UK, 1999.
Pawliszyn, J., *Applications of Solid Phase Microextraction*, The Royal Society of Chemistry, Cambridge, UK, 1999.
Dean, J. R., *Extraction Methods for Environmental Analysis*. Wiley, Chichester, UK, 1998.
Ramsey, E. D., *Analytical Supercritical Fluid Extraction Techniques*, Kluwer Academic Publishers, London, 1998.
Thurman, E. M. and Mills, M. S., *Solid Phase Extraction: Principles and Practice*, Wiley-Interscience, New York, 1998.
Pawliszyn, J., *Solid Phase Microextraction*, Wiley, New York, 1997.
Taylor, L. T., *Supercritical Fluid Extraction*, Wiley-Interscience, New York, 1996.
Berger, T. A., *Packed Column SFC*, The Royal Society of Chemistry, Cambridge, UK, 1995.
Brunner, G., *Gas Extraction*, Steinkopff Darmstadt-Springer, New York, 1994.
Luque de Castro, M. D., Valcarcel, M. and Tena, M. T., *Analytical Supercritical Fluid Extraction*, Springer-Verlag, New York, 1994.
McHugh, M. and Krukonis, V., *Supercritical Fluid Extraction*, Butterworths, Boston, MA, 1994.

Saito, M., Yamauchi, Y. and Okuyama, T., *Fractionation by Packed Column SFC and SFE*, VCH, New York, 1994.

Dean, J. R., *Applications of Supercritical Fluids in Industrial Analysis*, Blackie Academic and Professional, Glasgow, UK, 1993.

Bright, F. V. and McNally, M. E. P., *Supercritical Fluid Technology: Theoretical and Applied Approaches to Analytical Chemistry*, ACS Symposium Series 488, American Chemical Society, Washington, DC, 1992.

Jinno, K., *Hyphenated Techniques in Supercritical Fluid Chromatography and Extraction*, Elsevier, New York, 1992.

Wenclwaiak, B., *Analysis with Supercritical Fluids: Extraction and Chromatography*, Springer-Verlag, New York, 1992.

Westwood, S. A., *Supercritical Fluid Extraction and its Use in Chromatographic Sample Preparation*, Blackie Academic and Professional, Glasgow, UK 1992.

Lee, M. L. and Markides, K. E., *Analytical Supercritical Fluid Chromatography and Extraction*, Chromatography Conferences, Inc., Provo, UT, 1990.

Johnston, K. P. and Penninger, J. M. L., *Supercritical Fluid Science and Technology*, ACS Symposium Series 406, American Chemical Society, Washington, DC, 1989.

Charpentier, B. A. and Sevenants, M. R., *Supercritical Fluid Extraction and Chromatography*. ACS Symposium Series 366, American Chemical Society, Washington, DC, 1988.

Kingston, H. M. and Jassie, L. B., *Introduction to Microwave Sample Preparation*, ACS Professional Reference Book, American Chemical Society, Washington, DC, 1988.

Squires, T. G. and Paulaitis, M. E., *Supercritical Fluids*, ACS Symposium Series 329, American Chemical Society, Washington, DC, 1987.

Bock, R., *A Handbook of Decomposition Methods in Analytical Chemistry*, International Textbook Company, London, 1979.

12.2.2.7 Specific Books on Sampling

Baiuescu, G. E., Dumitrescu, P. and Gh. Zugravescu, P., *Sampling*, Ellis Horwood, London, 1991.

Keith, L. H., *Environmental Sampling and Analysis: A Practical Guide*, Lewis Publishers Inc., Chelsea, MI, 1991.

12.2.2.8 Specific Books on Speciation

Quevauviller, Ph., *Method Performance Studies for Speciation Analysis*, The Royal Society of Chemistry, Cambridge, UK, 1997.

Van der Sloot, H. A., Heasman, L. and Quevauviller, Ph., *Harmonization of Leaching/Extraction Tests*, Elsevier, Amsterdam, 1997.

Ure, A. M. and Davidson, C. M., *Chemical Speciation in the Environment*, Blackie Academic and Professional, Glasgow, UK, 1995.

Kramer, J. R. and Allen, H. E., *Metal Speciation: Theory, Analysis and Application*, Lewis Publishers, Chelsea, MI, 1988.

12.2.2.9 Specific Books on Statistics and Chemometrics

De Levie, R., *How to Use Excel in Analytical Chemistry and in General Scientific Data Analysis*, Cambridge University Press, Cambridge, UK, 2001.

Meier, P. C., *Statistical Methods in Analytical Chemistry*, 2nd Edn., Wiley, Chichester, UK, 2000.

Miller, J. N. and Miller, J. C., *Statistics and Chemometrics for Analytical Chemistry*, 4th Edn., Prentice Hall, Harlow, UK, 2000.

Farrant, T. J., *Practical Statistics for the Analytical Chemist*, The Royal Society of Chemistry, Cambridge, UK, 1997.

Massart, D. L., Vandeginste, B. G. M., Buydens, L. M. C., de Jong, S., Lewi, P. J. and Smeyers-Verbeke, J., *Handbook of Chemometrics and Qualimetrics: Part A*, Elsevier, Amsterdam, 1997.

Adams, M. J., *Chemometrics in Analytical Chemistry*, The Royal Society of Chemistry, Cambridge, UK, 1995.

12.2.3 Software

12.2.3.1 Safety

SoftCOSHH 2000, The Royal Society of Chemistry, Cambridge, UK, 2000.

12.2.3.2 Statistics and Chemometrics

Statistics for Analytical Chemists Softbook, The Royal Society of Chemistry, Cambridge, UK, 2001.

MultiSimplex®, Grabitech Solutions AB, Timra, Sweden (http://www.multisimplex.com).[†]

Statistica, 6.0, Statsoft, Tulsa, OK (http://www.statsoft.com).[†]

12.2.3.3 Analytical Techniques

AAS Softbook, The Royal Society of Chemistry, Cambridge, UK, 1995.

Advanced ICP Softbook, The Royal Society of Chemistry, Cambridge, UK, 1999.

GC Method Development Softbook, The Royal Society of Chemistry, Cambridge, UK, 1995.

GC Softbook, The Royal Society of Chemistry, Cambridge, UK, 1995.

HPLC Softbook, The Royal Society of Chemistry, Cambridge, UK, 1995.

Mass Spectrometry Softbook, The Royal Society of Chemistry, Cambridge, UK, 1995.

[†] As of June 2002. The products or material displayed are not endorsed by the author or the publisher.

12.2.4 CD-ROMs

The following are available from The Chemistry Video Consortium, Practical Laboratory Chemistry, Educational Media Film and Video Ltd, Harrow, UK:

- *Flame Photometry, AA and TGA Measurements* (using a flame photometer, using an atomic absorption spectrometer and thermogravimetric analysis).
- *Chromatographic Techniques* (TLC, column, ion-exchange and gas chromatography).
- *Microscale Chromatography* (TLC, column chromatography, gas chromatography and preparation of a Grignard reagent).
- *Electrochemical Techniques* (using galvanic cells, using conductometric cells, determining standard electrode potentials, determining solubility products, thermodynamic characteristics of cells, conductometric titrations and using an automatic titrator).
- *Polarimetry, Refractometry and Radiochemistry* (using a polarimeter, determining the refractive indices of liquids, measuring the rates of radioactive processes and measuring gas-phase emission spectra).
- *Inorganic Analysis* (gravimetric analysis).
- *Volumetric Techniques* (using a balance, using a pipette, using a burette and making up solutions).
- *Volumetric Analyses Methods* (doing a titration, some common end-points and potentiometric titrations).

12.2.5 Videos

Basic Laboratory Skills (LGC), The Royal Society of Chemistry, Cambridge, UK, 1998.
Further Laboratory Skills (LGC), The Royal Society of Chemistry, Cambridge, UK 1998.

12.2.6 Useful websites[†]

- *American Chemical Society*:
  http://www.acs.org
- *The Royal Society of Chemistry*:
  http://www.rsc.org
- *Society of Chemical Industry*:
  http://sci.mond.org

[†] As of June 2002. The products or material displayed are not endorsed by the author or the publisher.

- *International Union of Pure and Applied Chemistry (IUPAC)*:
  http://iupac.chemsoc.org
- *Laboratory of the Government Chemist (LGC)*:
  http://www.lgc.co.uk
- *United States Environmental Protection Agency (USEPA)*:
  http://www.epa.gov
- *National Institute of Standards and Technology (NIST) Laboratory*:
  http://www.cstl.nist.gov
- *National Institute of Standards and Technology (NIST) WebBook*:
  http://webbook.nist.gov
- *University of Sheffield*:
  http://www.chemdex.org
- *Specialized Information Services* – chemical information page on drugs, pesticides, environmental pollutants and other potential toxins:
  http://chem.sis.nlm.nih.gov
- *WebElements Periodic Table*:
  http://www.webelements.com

Summary

Accurate recording of experimental details prior to, during and after the practical work is an essential feature of all practical science. This chapter has attempted to capture the essential information that needs to be recorded at the same time as the practical work is being undertaken. While perhaps not totally comprehensive, it does offer the ability to be adapted to fit individual needs. In addition, a selected resource list is included to assist the reader.

Responses to Self-Assessment Questions

Chapter 1

Response 1.1

The following table shows the values that you should have obtained.

Quantity	m	μm	nm
6×10^{-7} m	0.000 000 6 m	0.6 μm	600 nm
Quantity	mol l^{-1}	mmol l^{-1}	μmol l^{-1}
2.5×10^{-3} mol l^{-1}	0.0025 mol l^{-1}	2.5 mmol l^{-1}	2500 μmol l^{-1}
Quantity	μg ml^{-1}	mg l^{-1}	ng μl^{-1}
8.75 ppm	8.75 μg ml^{-1}	8.75 mg l^{-1}	8.75 ng μl^{-1}

Response 1.2

You should have obtained the straight-line plot shown in Figure SAQ 1.2 below. This figure also shows the relevant mathematical relationship and R^2(R) value for such data (see text for further details).

Response 1.3

From Figure 1.1, it can be seen that the equation for the straight-line graph is as follows:

$$y = 0.0076x + 0.0004$$

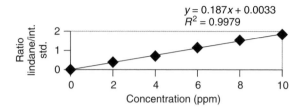

Figure SAQ 1.2 Analysis of lindane in waste water, using the direct calibration approach.

Thus, by rearranging this equation and inputting the y-value (0.026), it can be solved:

$$\frac{y}{0.0076} - 0.0004 = x$$

The concentration of lead from the graph is therefore 3.37 ppm (or 3.37 μg ml^{-1}).
   By using the dilution factor, you can now calculate the concentration of Pb in the original soil sample in the following way:

$$\frac{470\ ml}{g} \times \frac{3.37\ \mu g}{ml} = 1584\ \mu g\ g^{-1}$$

This result can also be expressed in terms of other units, e.g. 1584 mg kg^{-1} or 0.16 wt%.

Response 1.4

From Figure SAQ 1.2 it can been seen that the equation for the straight line graph is as follows:

$$y = 0.187x + 0.0033$$

Therefore, by rearranging this equation and inputting the y-value (0.26) it can be solved:

$$\frac{y}{0.187} - 0.0033 = x$$

The concentration of lindane from the graph is therefore 1.37 ppm (or 1.37 μg ml^{-1}).
   By using the dilution factor, you can now calculate the concentration of lindane in the waste water sample in the following way:

$$\frac{0.05\ ml}{ml} \times \frac{1.37\ \mu g}{ml} = 0.069\ \mu g\ ml^{-1}$$

This result can also be expressed in terms of other units, e.g. 69 ng ml^{-1} or 69 μg l^{-1}.

Chapter 2

Response 2.1

(a) The results would be described as accurate and precise.

(b) The results would be described as accurate but imprecise.

(c) The results would be described as inaccurate but precise.

(d) The results would be described as both inaccurate and imprecise.

Response 2.2

The graphs have been redrawn, indicating their linear working ranges (see Figure SAQ 2.2 below). Numerically, the linear working ranges are as follows:

(a) The linear working range is between 0 and 50 mg l^{-1}.

(b) The linear working range is between 0 and 10 mg l^{-1}.

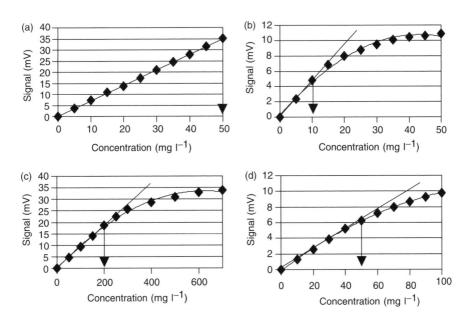

Figure SAQ 2.2 Plots showing different linear dynamic range values.

(c) The linear working range is between 0 and 200 mg l^{-1}.

(d) The linear working range is between 0 and 50 mg l^{-1}.

Chapter 3

Response 3.1

The magnitude of error, E, can be calculated as follows:

$$\sum(x_i - x)^2 = 466$$

$$S^2 = 466/9 = 51.8$$

$$V(\text{variance}) = 51.8/10 = 5.18$$

which gives:

$$E \text{ (at the 95\% confidence interval)} = \pm 2.26(5.18)^{0.5} = \pm 5.14 \text{ ppm}$$

(Note: the value of 2.26 was obtained from tables of critical values of Student's t statistics at the 95% confidence interval for $n - 1$.)[†]

This infers that in taking 10 samples, an error of 5.14 ppm was tolerated, and that the concentration of lead in the sample should be expressed as 93 ± 5.14 ppm.

Response 3.2

The number of samples required can be calculated by using equation (3.4) as follows:

$$(1.96)^2(51.8)/2^2 = 49.7$$

Therefore, 50 samples are required.

(Note: the value of 1.96 was obtained from tables of critical values of Student's t statistics at the 95% confidence interval for $n = \infty$.)[†]

Chapter 4

Response 4.1

In either case, you would first need to ensure that you have the necessary container for the storage of the sample and that the container has been pre-treated

---

[†] Student's t statistics (distributions) are widely used in solving statistical problems in chemical analysis involving small numbers of samples ($n < 30$). For further details, see the texts listed in Section 12.2.2.9.

appropriately. In addition, care will have been needed to obtain a representative sample of the water (see Chapter 3) and in an appropriate manner.

It is also good practice when sampling water to rinse out the pre-cleaned container with the sample prior to sample collection. This allows the container to be pre-conditioned with the sample prior to collection.

Total lead analysis. It is recommended that 100 ml of the water sample is introduced into a pre-cleaned and rinsed polyethylene container with a polypropylene cap, or into a glass bottle. In both cases, acidify the water to pH < 2 with nitric acid. Ensure that the container is completely full of the sample. In this situation, the sample can be held for up to six months for the analysis of total lead.

Sulfate analysis. It is recommended that 50 ml of the water sample is introduced into a pre-cleaned and rinsed polyethylene container with a polypropylene cap, or into a glass bottle. In both cases, store the sample at $4°C$, either in a cool box on-site and during transportation, and then in a fridge in the laboratory. Ensure that the container is completely full of the sample. In this situation, the sample can be held for up to 28 days for the analysis of sulfate.

Dieldrin analysis. Dieldrin is an organochlorine pesticide. It is recommended that 500 ml of the water sample is introduced into a pre-cleaned and rinsed glass bottle. For preservation of the sample, two options are possible, i.e. either add 1 ml of a 10 mg ml^{-1} $HgCl_2$ solution, or add the appropriate extraction solvent (or pre-concentrate the sample on a solid-phase extraction cartridge (see Chapter 8)). Ensure that the container is completely full of the sample. In this situation, the sample can be held for up to seven days if using $HgCl_2$ solution (or 40 days, if extraction solvent is added) for the analysis of dieldrin.

Chapter 5

Response 5.1

Hydrofluoric acid is the reagent used for dissolving silica-based materials. The silicates are converted to a more volatile species in solution, according to the following equation:

$$SiO_2 + 6HF = H_2(SiF_6) + 2H_2SO_4$$

Response 5.2

While a different mass of sample is taken, ranging from 50–2000 mg for fish and 25–2000 mg for sediment, most methods involve some form of liquid–solid extraction, i.e. a known mass of sample is extracted with a solvent. Often, the sample has previously been acidified and the solvent is usually toluene. Then, depending upon the separation technique, some additional sample work-up is required. Gas chromatography appears to be the most common method of separation. Often, for gas chromatography, the methylmercury requires to be converted into a volatile form. This has been most commonly camed out with either $NaBH_4$ or $NaBEt_4$, or by butylation with a Grignard reagent (butylmagnesium chloride). A range of detection systems have been used for gas chromatography, including the conventional type of detector, i.e. electron-capture detection, through to atomic spectroscopy, e.g. cold-vapour atomic absorption spectroscopy.

Response 5.3

While a different mass of sample is taken, ranging from 0.5–25 g, most methods involve some form of liquid–solid extraction, i.e. a known mass of sample is extracted with a solvent. A range of solvents have been used, including acetic acid. The most common separation technique is gas chromatography, with this requiring the conversion of butyltin into a volatile form. This has been most commonly camed out with either $NaBH_4$ or $NaBEt_4$, or with a Grignard reagent (butyl-, ethyl- or pentylmagnesium chloride). A range of detection systems have been used for gas chromatography, such as mass spectrometry or flame photometric detection, as well as atomic spectroscopy, e.g. quartz-furnace atomic absorption spectroscopy.

Response 5.4

While a different mass of sample is taken, ranging from 0.15–2 g, most methods involve ultrasonic extraction, i.e. a known mass of sample is extracted with a solvent by using ultrasonic agitation. The most common solvent system is methanol and water. The most commonly used separation technique is high performance liquid chromatography, with anion-exchange being the most popular. Atomic spectroscopy, coupled with high performance liquid chromatography, was the method of choice for this analysis.

Chapter 6

Response 6.1

By changing the pH of the solution it is (potentially) possible to achieve selectivity between Sn and Pb. Tin can only form an APDC complex in the pH range 2–8, while Pb can do the same thing between pH 2–14 (see Table 6.1). Therefore, if the Pb complex was formed at pH values >8, no similar complex for Sn should be formed. Thus, selectivity between Pb and Sn should be possible.

Chapter 7

Response 7.1

First of all, it is important to identify what the acronyms represent. The following have been used: aMAE, atmospheric microwave-assisted extraction; SFE, supercritical fluid extraction; pMAE, pressurized microwave-assisted extraction; PFE, pressurized fluid extraction; ASE, accelerated solvent extraction; MSPD, matrix solid-phase dispersion.

The *conventional* techniques are Soxhlet, Soxtec, ultrasonic and shake-flask, while the *instrumental* techniques are aMAE, SFE, pMAE and PFE (ASE).

Matrix solid-phase dispersion (MSPD) is a relatively new technique, with, as yet, only a limited number of applications.

Response 7.2

A typical extraction efficiency–time profile is shown below in Figure SAQ 7.2. As the extraction time is increased, a greater recovery of analyte is achieved. However, the maximum recovery is achieved in the first time-domain (0 to 0.25t); thereafter, smaller quantities of analyte are recovered. In the end (t), it may not be possible to remove all the analyte from the matrix. In practice, therefore, a compromise extraction time is often chosen.

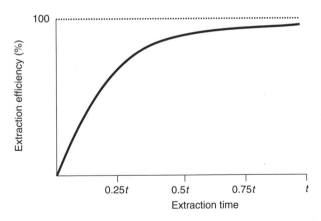

Figure SAQ 7.2 A typical extraction efficiency–time profile for the removal of an analyte from a matrix.

Response 7.3

A comprehensive evaluation of the various techniques, based on the suggested guidelines, is presented below in Table SAQ 7.3. Hopefully, your own survey will have contained most of the details given in this table.

Table SAQ 7.3 Comparison of the various techniques used for the extraction of organic analytes from solid matrices[a]

Feature/conditions	Soxhlet	Soxtec	Ultrasonic	MSPD	SFE	MAE	PFE
Description of method	Utilizes cooled condensed solvents passing over the sample contained in a thimble to extract analytes. Uses specialist glassware and heating apparatus	Also known as automated Soxhlet. Soxtec places the sample into the boiling solvent and then flushes clean solvent over it. Faster than Soxhlet	Sample is covered with organic solvent, and then a sonic horn is placed inside the container with the solvent and sample	Sample is mixed with a dispersant, e.g. C18 media, and then placed in an empty solid-phase extraction cartridge. Analytes eluted with appropriate solvent	Utilizes supercritical CO_2, with or without organic modifier, to extract analytes. Pressures up to 680 atm and temperature up to 250°C can be used. Analytes collected in solvent	Utilizes microwave radiation to heat solvent. Can be carried out either at elevated pressures or atmospheric pressure	Utilizes high temperature (100°C) and pressure (2000 psi) to extract analytes. Solvent and analytes are flushed from the extraction vessel by using a small volume of fresh solvent and a N_2 purge. Fully automated
Sample mass (g)	10	10	1–5	1–5	1–10	2–5	Up to 30 g
Extraction time	6, 12 or 24 h	Reduced time compared to Soxhlet, i.e. 2–4 h	Typically 5–15 min, but repeated for up to three times	Should be possible in under 30 min	30 min–1 h	20 min (plus 30 min cooling and pressure reduction)	12 min
Solvent type	Acetone:hexane (1:1 (vol/vol)); acetone:DCM (1:1(vol/vol)); DCM only; toluene:methanol (10:1(vol/vol)); alternatively, a solvent system of your own choice	As Soxhlet	As Soxhlet	Requires optimization	CO_2 (plus organic modifier). Tetrachloroethene used as the collection solvent for TPHs for determination by FTIR; otherwise, DCM	Typically, acetone:hexane (1:1(vol/vol)). The solvent(s) is/are required to be able to absorb microwave energy	Acetone:hexane (1:1(vol/vol)) or acetone; DCM (1:1(vol/vol)) for OCPs, semi-volatile organics, PCBs or OPPs; acetone: DCM:phosphoric acid (250:125:15(vol/vol)) for chlorinated herbicides

Table SAQ 7.3 (*continued*)

Feature/ conditions	Soxhlet	Soxtec	Ultrasonic	MSPD	SFE	MAE	PFE
Solvent consumption (ml) per extraction	150–300	40–50	50–100	20–50	10–20	25–45	25
Sequential or simultaneous	sequential (but multiple assemblies can operate simultaneously)	Systems available for two or six samples simultaneously	Sequential	Sequential	Sequential	Simultaneous (up to 14 vessels can be extracted simultaneously)	Sequential
Equipment cost	Low	Low–moderate	Low–moderate	Low–moderate	High	Moderate	High
Level of automation	Minimal	Minimal	Minimal	Minimal	Minimal to high	Minimal	Fully automated – up to 24 samples can be extracted
EPA method	3540	3541	3550	None	3560 for TPHs, 3561 for PAHs and 3562 for PCBs and OCPs	3546	3545

a DCM, dichloromethane; FTIR, Fourier-transform infrared (spectroscopy); TPHs, total petroleum hydrocarbons; PAHs, polycyclic aromatic hydrocarbons; OCPs, organochlorine pesticides; PCBs, polychlorinated biphenyl; OPPs, organophosphate pesticides.

Chapter 8

Response 8.1

Suggesting a procedure for the solid-phase extraction of any of the compounds mentioned is not straightforward. It is always a good starting point to consider the main variables in any SPE procedure. These include the following: nature of the analyte, sorbent to be used, wetting/conditioning of the sorbent, loading the sample, and analyte elution conditions. In considering the responses to the above problems, some examples from the literature have been chosen as illustrations.

(a) Organochlorine pesticides from drinking and surface water [5]
   Analyte: organochlorine pesticides.
   Sorbent: C18 cartridge.
   Wetting/conditioning the sorbent: 2×5 ml of methanol, followed by 2×2 ml of water.
   Loading: 1 l volumes of drinking and surface water samples from the Gdansk district, Poland, were passed at a flow rate of approximately $10\ \mathrm{ml\,min^{-1}}$.
   Elution: 2×2.5 ml of *n*-pentane and dichloromethane (1:1(vol/vol)).
   Other comments: Both extracts were combined and evaporated to 0.5 ml, prior to GC–ECD analysis.

(b) Carbamate pesticides from river water [6]
   Analyte: 10 carbamate pesticides.
   Sorbent: C18, 'Empore' discs.
   Wetting/conditioning the sorbent: 10 ml of methanol under vacuum, followed by 10 ml of acetonitrile (after drying). Subsequently, 30 ml of water, ensuring that the disc does not become dry prior to addition of sample.
   Loading: 2 l of water from the River Erbo (Spain).
   Elution: After drying for 1 h under vacuum, elution was carried out by using 2×10 ml of acetonitrile.
   Other comments: Extract was evaporated to dryness and the residue diluted in 500 µl of methanol prior to analysis by HPLC–MS.

(c) Acid herbicides from surface water [7]
   Analyte: 8 polar acid herbicides.
   Sorbent: C18 cartridge.
   Wetting/conditioning the sorbent: 10 ml of methanol, followed by 10 ml of deionized water acidified to pH 2.4–2.6 with HCl. The cartridge was not allowed to run dry during this process.
   Loading: 500 ml of a surface water sample, acidified to pH 2.4–2.6 with 2 M HCl, and containing 5 ml of methanol and 1.0 ml of internal standard, was percolated through the cartridge under vacuum at a rate of approximately $15\ \mathrm{ml\,min^{-1}}$. Cartridge was then air-dried under vacuum for 5 min and centrifuged for 15 min at $500g$ to remove the remaining water.

Rinsing: The cartridge was then washed with 0.75 ml of methanol and the solution discarded.

Elution: 2×2 ml of methanol, under vacuum, at a rate of approximately 2 ml min^{-1}.

Other comments: Extract was then adjusted to 10 ml with methanol, followed by derivatization and column chromatographic clean-up, prior to GC analysis.

Response 8.2

A detailed evaluation of the various techniques, based on the suggested guidelines, is presented below in Table SAQ 8.2. Hopefully, your own survey will have contained most of the information given in this table.

Table SAQ 8.2 Comparison of the various techniques used for the separation and/or pre-concentration of organic compounds in solution

Features/conditions	Liquid–liquid	SPE	SPME
Description of method	Sample is partitioned between two immiscible solvents; continuous and discontinuous operation possible	Analyte retained on a solid absorbent; extraneous sample material washed from sorbent. Desorption of analyte using organic solvent	Analyte retained on a sorbent-containing fibre attached to a silica support. Fibre protected by a syringe barrel when not in use. Most commonly found for GC applications
Sample size	1–2 l	1–1000 ml	1–1000 ml
Extraction time	Discontinuous, 20 min; continuous, up to 24 h	10–20 min	10–60 min (requires optimization)
Solvent consumption (ml) per extraction	30–60 ml for discontinuous; up to 500 ml for continuous	Organic solvent required for wetting sorbent and elution of analyte (10–20 ml)	No solvent required
Equipment cost	Low	Low–high (depends on degree of automation)	Low (but also available as an automated system)
Acceptability	Wide acceptance for isolating organic compounds	Widely acceptable	Gaining in popularity; new technology
EPA method	3510 and 3520	3535	None

Chapter 11

Response 11.1

To answer this question requires knowledge of what the actual analytical techniques measure, plus knowledge of what the pollutants are. If you are unsure about what each analytical technique measures, then complete this chapter before attempting to answer this question. Knowledge of the pollutants may have been gained while working through the book to this stage. If not, you should consult the 'Resources' section towards the end of this book (see Section 12.2).

Attempts to determine the appropriate analytical technique for each pollutant can be carried out at three different knowledge levels. At the first level, the 'novice' level, it should be easy to establish whether the pollutant can be determined by using an analytical technique capable of determining organic or inorganic pollutants. However, with a greater understanding of the analytical techniques, and what they are capable of, comes additional questions, e.g. at what concentration level is the pollutant likely to be present? Is the pollutant in an aqueous solution or in solid form? This additional level of knowledge, or 'intermediate' level, comes with experience and a more advanced study of the various techniques. I would define the 'advanced' level as appropriate for someone with both theoretical and detailed practical knowledge of the techniques.

The completed table below indicates the level of knowledge expected for someone who has some basic understanding of the various analytical techniques (the 'novice' level).

Pollutant	Technique
Polycyclic aromatic hydrocarbons (PAHs)	GC or HPLC
Lead	FAAS/ICP–AES/ICP–MS
Phenols	GC or HPLC
Chlorinated pesticides	GC
Arsenic	FAAS/ICP–AES/ICP–MS
Chromium(VI)	FAAS/ICP–AES/ICP–MS
Benzene–toluene–ethylbenzene–xylene(s) (BTEX)	GC
Total petroleum hydrocarbons (TPHs)	GC
Nitrate	IC
Sodium	FP

However, for students with an additional level of knowledge 'intermediate' level, the choice of analytical technique has some additional complexities (as discussed below).

Polycyclic aromatic hydrocarbons (PAHs). These can be routinely determined by using GC with FID or MSD. For trace level analysis, the use of MSD is preferred as it allows the mass spectrometer to be used in the selected-ion monitoring (SIM) mode, which has the additional benefit of providing target analysis of the selected PAHs via their ions (mass/charge ratio). However, it is also possible to determine PAHs with UV/Vis or fluorescence detection. As all of the PAHs are aromatic (i.e. possess a chromophore), they can be detected in both of these ways. The latter detection method offers additional sensitivity when compared to UV/vis detection.

Lead. This inorganic pollutant can be determined by a whole range of analytical techniques; in addition, information is often required on the concentration level of lead. In terms of sensitivity (least sensitive first), the following is suggested: FAAS < ICP–AES < GFAAS = ASV < ICP–MS. While all of these techniques can determine lead in solution, the use of XRF allows lead to be determined directly in the solid.

Phenols. These can be determined by both GC and HPLC. In this situation, the choice of technique will be dependent upon other factors, such as the choice of separation column. In some cases, phenols can only be analysed by GC after prior derivatization of the phenols into more volatile species. The preferred method is therefore suggested to be HPLC with UV/Vis detection.

Chlorinated pesticides. These can be determined by GC. The specific description of the pollutants as chlorinated pesticides should immediately indicate that the choice of detection should be ECD. While this method will provide specific and sensitive detection of chlorinated pesticides, the optional benefits of MSD (see PAHs) may be advantageous.

Arsenic. This is a volatile metalloid capable of forming an hydride. The immediate choice, therefore, would be HyFAAS. This would provide a sensitive method of determining arsenic. In addition, the use of inductively coupled plasma techniques is also a possibility. It would not be unreasonable for maximum sensitivity to couple the hydride generation apparatus up to the ICP–MS system. However, the use of the latter has one area of concern with respect to arsenic determination. As arsenic is monoisotopic (only one isotope, at 75 amu), interferences can result in the presence of chlorine-forming species e.g. HCl. If this is the case, the use of chlorine-containing material should be avoided. Alternatively, remove the potential interference from chlorine-containing material by performing some chromatographic separation prior to the ICP–MS stage.

Chromium(VI). This inorganic pollutant can be determined by a whole range of analytical techniques, e.g. FAAS, ICP–AES/MS or XRF. However, in this situation the requirement is to determine a particular oxidation state of chromium. While not mentioned specifically in the list of analytical techniques above, a method does exist for the spectrophotometric determination of hexavalent chromium based on extraction with a chelating agent (see Chapter 5 for details).

Alternative analytical techniques could also be employed. For example, ion chromatography could be used for separation, followed by either spectrophotometric detection after post-column chelation, or ICP–AES/MS detection.

Benzene–toluene–ethylbenzene–xylene(s) (BTEX). In this situation, the method of choice would be GC–FID. However, as discussed previously, the optional benefits of an mass-selective detection (see PAHs) may be advantageous.

Total petroleum hydrocarbons (TPHs). A method for the determination of TPHs could be GC–FID. As before, the optional benefits of MSD (see PAHs) may be advantageous. However, as information is only required on total petroleum hydrocarbons, an alternative analytical technique is possible, i.e. FTIR spectroscopy. In this situation, C–H bands are observed in the region 2800–3000 cm^{-1}. Care, however, must be taken in the choice of the solvent used for the preparation of standards and samples. The solvent must not contain carbon and hydrogen. A typical solvent to use, therefore, would be tetrachloroethylene.

Nitrate. In this case, the method of choice would be IC, with conductivity detection. The use of a dedicated ion chromatograph with an anion-exchange column allows the separation of nitrate (plus fluoride, chloride, bromide, nitrite, sulfate, etc.).

Sodium. In this situation, the method of choice would be FP.

Glossary of Terms

This section contains a glossary of terms, all of which are used in the text. It is not intended to be exhaustive, but to explain briefly those terms which often cause difficulties or may be confusing to the inexperienced reader.

Absorbance   A measurement of the absorption of light; absorbance $= \log I_0/I$, where I_0 and I are the intensities of the incident and transmitted radiation, respectively.

Accelerated solvent extraction (ASE)   Method of extracting analytes from matrices using a solvent at elevated pressures and temperatures (*see also* Pressurized fluid extraction).

Accuracy   A quantity referring to the difference between the mean of a set of results or an individual result and the value which is accepted as the true or correct value for the quantity measured.

Acid digestion   Use of acid (and often heat) to destroy the organic matrix of a sample to liberate the metal content.

Aliquot   A known amount of a homogenous material assumed to be taken with negligible sampling error.

Analyte   The component of a sample which is ultimately determined directly or indirectly.

Anion   Ion having a negative charge; an atom with extra electrons. Atoms of non-metals, in solution, become anions.

Bias   Characterizes the systematic error in a given analytical procedure and is the (positive or negative) deviation of the mean analytical result from the (known or assumed) true value.

Blow-down   Removal of liquids and/or solids from a vessel by the use of pressure; often used to remove solvents to pre-concentrate the analyte.

BTEX   Acronym used to describe a mixture of the following volatile organic compounds: benzene, toluene, ethylbenzene and *ortho-*, *meta-* and *para-*xylenes.

Calibration   The set of operations which establish, under specified conditions, the relationship between values indicated by a measuring instrument or measuring system and the corresponding known values of the measurand.

Calibration curve   Graphical representation of the measuring signal (response) as a function of the quantity of analyte.

Cation   Ion having a positive charge. Atoms of metals, in solution, become cations.

Certified Reference Material (CRM)   Reference material, accompanied by a certificate, one or more of whose property values are certified by a procedure which establishes its traceability to an accurate realization of the unit in which the property values are expressed, and for which each certified value is accompanied by an uncertainty at a stated level of confidence.

Complexing agent   The chemical species (an ion or a compound) which will bond to a metal ion using lone pairs of electrons.

Confidence interval   Range of values which contains the true value at a given level of probability. The level of probability is called the *confidence level.*

Confidence limit   The extreme values or end values in a confidence interval.

Contamination   In trace analysis, this is the unintentional introduction of analyte(s) or other species which are not present in the original sample and which may cause an error in the determination. It can occur at any stage in the analysis. Quality assurance procedures, such as the analyses of blanks or reference materials, are used to check for contamination problems.

Control of Substances Hazardous to Health (COSHH)   Regulations that impose specific legal requirements for risk assessment wherever hazardous chemicals or biological agents are used.

Co-precipitation   The inclusion of otherwise soluble ions during the precipitation of lower-solubility species.

Degradation   The breakdown of organic molecules into simpler species through a number of distinct stages. This may be by either chemical or biological means.

Dilution factor   The mathematical factor applied to the determined value (data obtained from a calibration graph) that allows the concentration in the original sample to be determined. Frequently, for solid samples this will involve a sample weight and a volume to which the digested/extracted sample is made up to, prior to analysis. For liquid samples, this will involve an initial sample volume and a volume to which the digested/extracted sample is made up to, prior to analysis.

Dissolved material   Refers to material which will pass through a 0.45 μm membrane filter assembly prior to sample acidification.

Dry ashing   Use of heat to destroy the organic matrix of a sample to liberate the metal content.

Eluent   The mobile liquid phase in liquid or in solid-phase extraction.

Error   The error of an analytical result is the difference between the result and a 'true' value.

   Random error   Result of a measurement minus the mean that would result from an infinite number of measurements of the same measurand carried out under repeatability conditions.

   Systematic error   Mean that would result from an infinite number of measurements of the same measurand carried out under repeatability conditions minus the true value of the measurand.

Extraction   The removal of a soluble material from a solid mixture by means of a solvent, or the removal of one or more components from a liquid mixture by use of a solvent with which the liquid is immiscible or nearly so.

Figure of merit   A parameter that describes the quality of performance of an instrument or an analytical procedure.

Fitness for purpose   The degree to which data produced by a measurement process enables a user to make technically and administratively correct decisions for a stated purpose.

Heterogeneity   The degree to which a property or a constituent is randomly distributed throughout a quantity of material. The degree of heterogeneity is the determining factor of sampling error.

Homogeneity   The degree to which a property or a constituent is uniformly distributed throughout a quantity of material. A material may be homogenous with respect to one analyte but heterogeneous with respect to another.

Humic acid   Naturally occurring high-molecular-mass organic compounds which are acid-soluble but are precipitated by base.

Interferent   Any component of the sample affecting the final measurement.

Kuderna–Danish evaporator   Apparatus for sample concentration, consisting of a small (10 ml) graduated test-tube connected directly beneath a 250 or 500 ml flask. A steam bath provides heat for evaporation with the concentrate collecting in the test-tube.

Leachate   The liquid after passing through a substance which contains soluble extracts.

Limit of detection   The detection limit of an individual analytical procedure is the lowest amount of an analyte in a sample which can be detected but not necessarily quantified as an exact value. The limit of detection, expressed as the concentration c_L or the quantity q_L, is derived from the smallest measure, x_L, that can be detected with reasonable certainty for a given procedure. The value x_L is given by the equation:

$$x_L = x_{bl} + k s_{bl}$$

where x_{bl} is the mean of the blank measures, s_{bl} is the standard deviation of the blank measures and k is a numerical factor chosen according to the confidence level required. For many purposes, the limit of detection is taken to be $3s_{bl}$ or $3 \times$ 'the signal-to-noise ratio', assuming a zero blank.

Limit of quantitation   For an individual analytical procedure, this is the lowest amount of an analyte in a sample which can be quantitatively determined with suitable uncertainty. It may also be referred to as the *limit of determination*. The limit of quantitation can be taken as $10 \times$ 'the signal-to-noise ratio', assuming a zero blank.

Linear dynamic range (LDR)   The concentration range over which the analytical working calibration curve remains linear.

Linearity   Defines the ability of the method to obtain test results proportional to the concentration of analyte.

Liquid–liquid extraction   A method of extracting a desired component from a liquid mixture by bringing the solution into contact with a second liquid, the solvent, in which the component is also soluble, and which is immiscible with the first liquid, or nearly so.

Matrix   The carrier of the test component (analyte); all of the constituents of the material except the analyte, or the material with as low a concentration of the analyte as it is possible to obtain.

Measurand   Particular quantity subject to measurement.

Method   The overall, systematic procedure required to undertake an analysis. This includes all stages of the analysis – not just the (instrumental) end determination.

Microwave-assisted extraction (MAE)   Method of extracting analytes from matrices using solvent at elevated temperatures (and pressures) based on microwave radiation. Can be carried out in either open or sealed vessels.

Microwave digestion   Method of digesting an organic matrix to liberate metal content using acid at elevated temperatures (and pressures) based on microwave radiation. Can be carried out in either open or sealed vessels.

Organometallic   An organic compound in which a metal is covalently bonded to carbon.

Outlier   An observation in a set of data that appears to be inconsistent with the remainder of that set.

Pesticide   Any substance or mixture of substances intended for preventing, destroying, repelling, or mitigating any pest. The latter can be insects, mice and other animals, unwanted plants (weeds), fungi, or microorganisms such as bacteria and viruses. Although often misunderstood to refer only to *insecticides*, the term 'pesticide' also applies to herbicides, fungicides, and various other substances used to control pests.

Polycyclic aromatic hydrocarbons (PAHs)   These are a large group of organic compounds, comprising two or more aromatic rings, which are widely distributed in the environment.

Precision   The closeness of agreement between independent test results obtained under stipulated conditions.

Pressurized fluid extraction (PFE)   Method of extracting analytes from matrices using solvent at elevated pressures and temperatures (*see also* Accelerated solvent extraction).

Qualitative analysis   Chemical analysis designed to identify the components of a substance or mixture.

Quality assurance   All of those planned and systematic actions necessary to provide adequate confidence that a product or services will satisfy given requirements for quality.

Quality control   The operational techniques and activities that are used to fulfil requirements of quality.

Quality control chart   A graphical record of the monitoring of control samples which helps to determine the reliability of the results.

Quantitative analysis   Chemical analysis which is normally taken to mean the numerical measurement of one or more analytes to the required level of confidence.

Reagent   A test substance which is added to a system in order to bring about a reaction or to see whether a reaction occurs (e.g. an analytical reagent).

Reagent blank   A solution obtained by carrying out all of the steps of an analytical procedure in the absence of a sample.

Recovery   The fraction of the total quantity of a substance recoverable following a chemical procedure.

Reference material   Substance or material, one or more of whose property values are sufficiently homogeneous and well established to be used for the calibration of an apparatus, the assessment of a measurement method, or for assigning values to materials.

Repeatability   Precision under repeatability conditions, i.e. conditions where independent test results are obtained with the same method on identical test items in the same laboratory, by the same operator, using the same equipment within short intervals of time.

Reproducibility   Precision under reproducibility conditions, i.e. conditions where test results are obtained with the same method on identical test items in different laboratories, with different operators, using different equipment.

Robustness   A measure of the capacity of an analytical procedure to remain unaffected by small, but deliberate variations in method parameters, and which provides an indication of its reliability during normal usage. Sometimes referred to as *ruggedness*.

Rotary evaporation   Removal of solvents by distillation under vacuum.

Sample   A portion of material selected from a larger quantity of material. The term needs to be qualified, e.g. representative sample, sub-sample, etc.

Selectivity (in analysis)   (i) Qualitative – the extent to which other substances interfere with the determination of a substance according to a given procedure.

(ii) Quantitative – a term used in conjunction with another substantive (e.g. constant, coefficient, index, factor, number, etc.) for the quantitative characterization of interferences.

Sensitivity   The change in the response of a measuring instrument divided by the corresponding change in stimulus.

Signal-to-noise ratio   A measure of the relative influence of noise on a control signal. Usually taken as the magnitude of the signal divided by the standard deviation of the background signal.

Shake-flask extraction   Method of extracting analytes from matrices using agitation or shaking in the presence of a solvent.

Solid-phase extraction (SPE)   A sample preparation technique that uses a solid-phase packing contained in a small plastic cartridge. The solid stationary phases are the same as HPLC packings; however, the principle is different from HPLC. The process, as most often practised, requires four steps: conditioning the sorbent, adding the sample, washing away the impurities, and eluting the sample in as small a volume as possible with a strong solvent.

Solid-phase microextraction (SPME)   A sample preparation technique that uses a fused silica fibre coated with a polymeric phase to sample either an aqueous solution or the headspace above a sample. Analytes are absorbed by the polymer coating and the SPME fibre is directly transferred to a GC injector or special HPLC injector for desorption and analysis.

Solvent extraction   The removal of a soluble material from a solid mixture by means of a solvent, or the removal of one or more components from a liquid mixture by use of a solvent with which the liquid is immiscible or nearly so.

Soxhlet extraction   Equipment used for the continuous extraction of a solid by a solvent. The material to be extracted is placed in a porous cellulose thimble, and continually condensing solvent is then allowed to percolate through it, and return to the boiling vessel, either continuously or intermittently.

Speciation   The process of identifying and quantifying the different defined species, forms or phases present in a material, or the description of the amounts and types of these species, forms or phases present.

Specificity   The ability of a method to measure only what it is intended to measure – the ability to assess unequivocally the analyte in the presence of components which may be expected to be present. Typically, these might include impurities, degradants, matrices, etc.

Spiked sample   'Spiking a sample' is a widely used term taken to mean the addition of a known quantity of analyte to a matrix which is close to or identical with that of the sample(s) of interest.

Standard (general)   An entity established by consensus and approved by a recognized body. It may refer to a material or solution (e.g. an organic compound of known purity or an aqueous solution of a metal of agreed concentration), or a document (e.g. a methodology for an analysis or a quality system). The relevant terms are:

Analytical standard (*also known as* **Standard solution**) A solution or matrix containing the analyte which will be used to check the performance of the method/instrument.

Calibration standard The solution or matrix containing the analyte (measurand) at a known value with which to establish a corresponding response from the method/instrument.

Internal standard A measurand, similar to but not identical with the analyte, which is combined with the sample.

External standard A measurand, usually identical with the analyte, which is analysed separately from the sample.

Standard method A procedure for carrying out a chemical analysis which has been documented and approved by a recognized body.

Standard addition The addition of a known amount of analyte to the sample in order to determine the relative response of the detector to the analyte within the sample matrix. The relative response is then used to assess the sample analyte concentration.

Stock solution This is generally a standard or reagent solution of known accepted stability, which has been prepared in relatively large amounts, of which portions are used as required. Frequently, such portions are used following further dilution.

Sub-sample This may be (i) a portion of the sample obtained by selection or division, (ii) an individual unit of the lot taken as part of the sample, or (iii) the final unit of multi-stage sampling.

Supercritical fluid extraction (SFE) A method of extracting analytes from matrices using a supercritical fluid at elevated pressures and temperatures. The term *supercritical fluid* is used to describe any substance above its critical temperature and critical pressure.

True value A value consistent with the definition of a given particular quantity.

Ultrasonic extraction A method of extracting analytes from matrices with solvent, using either an ultrasonic bath or probe.

Uncertainty Parameter associated with the result of a measurement, which characterizes the dispersion of the values that could reasonably be attributed to the measurand.

SI Units and Physical Constants

SI Units

The SI system of units is generally used throughout this book. It should be noted, however, that according to present practice, there are some exceptions to this, for example, wavenumber (cm^{-1}) and ionization energy (eV).

Base SI units and physical quantities

Quantity	Symbol	SI Unit	Symbol
length	l	metre	m
mass	m	kilogram	kg
time	t	second	s
electric current	I	ampere	A
thermodynamic temperature	T	kelvin	K
amount of substance	n	mole	mol
luminous intensity	I_v	candela	cd

Prefixes used for SI units

Factor	Prefix	Symbol
10^{21}	zetta	Z
10^{18}	exa	E
10^{15}	peta	P
10^{12}	tera	T
10^{9}	giga	G
10^{6}	mega	M
10^{3}	kilo	k

(*continued overleaf*)

Prefixes used for SI units (*continued*)

Factor	Prefix	Symbol
10^2	hecto	h
10	deca	da
10^{-1}	deci	d
10^{-2}	centi	c
10^{-3}	milli	m
10^{-6}	micro	μ
10^{-9}	nano	n
10^{-12}	pico	p
10^{-15}	femto	f
10^{-18}	atto	a
10^{-21}	zepto	z

Derived SI units with special names and symbols

Physical quantity	SI unit		Expression in terms of base or derived SI units
	Name	Symbol	
frequency	hertz	Hz	$1\ \text{Hz} = 1\ \text{s}^{-1}$
force	newton	N	$1\ \text{N} = 1\ \text{kg}\,\text{m}\,\text{s}^{-2}$
pressure; stress	pascal	Pa	$1\ \text{Pa} = 1\ \text{Nm}^{-2}$
energy; work; quantity of heat	joule	J	$1\ \text{J} = 1\ \text{Nm}$
power	watt	W	$1\ \text{W} = 1\ \text{J}\,\text{s}^{-1}$
electric charge; quantity of electricity	coulomb	C	$1\ \text{C} = 1\ \text{A}\,\text{s}$
electric potential; potential difference; electromotive force; tension	volt	V	$1\ \text{V} = 1\ \text{J}\,\text{C}^{-1}$
electric capacitance	farad	F	$1\ \text{F} = 1\ \text{C}\,\text{V}^{-1}$
electric resistance	ohm	Ω	$1\ \Omega = 1\ \text{V}\,\text{A}^{-1}$
electric conductance	siemens	S	$1\ \text{S} = 1\ \Omega^{-1}$
magnetic flux; flux of magnetic induction	weber	Wb	$1\ \text{Wb} = 1\ \text{V}\,\text{s}$
magnetic flux density; magnetic induction	tesla	T	$1\ \text{T} = 1\ \text{Wb}\,\text{m}^{-2}$
inductance	henry	H	$1\ \text{H} = 1\ \text{Wb}\,\text{A}^{-1}$
Celsius temperature	degree Celsius	°C	$1\,°\text{C} = 1\ \text{K}$
luminous flux	lumen	lm	$1\ \text{lm} = 1\ \text{cd}\,\text{sr}$

Derived SI units with special names and symbols (*continued*)

Physical quantity	SI unit		Expression in terms of base or derived SI units
	Name	Symbol	
illuminance	lux	lx	$1 \, \text{lx} = 1 \, \text{lm} \, \text{m}^{-2}$
activity (of a radionuclide)	becquerel	Bq	$1 \, \text{Bq} = 1 \, \text{s}^{-1}$
absorbed dose; specific energy	gray	Gy	$1 \, \text{Gy} = 1 \, \text{J} \, \text{kg}^{-1}$
dose equivalent	sievert	Sv	$1 \, \text{Sv} = 1 \, \text{J} \, \text{kg}^{-1}$
plane angle	radian	rad	1^a
solid angle	steradian	sr	1^a

[a] rad and sr may be included or omitted in expressions for the derived units.

Physical Constants

Recommended values of selected physical constants[a]

Constant	Symbol	Value
acceleration of free fall (acceleration due to gravity)	g_n	$9.806\,65 \, \text{m} \, \text{s}^{-2}$ [b]
atomic mass constant (unified atomic mass unit)	m_u	$1.660\,540\,2(10) \times 10^{-27} \, \text{kg}$
Avogadro constant	L, N_A	$6.022\,136\,7(36) \times 10^{23} \, \text{mol}^{-1}$
Boltzmann constant	k_B	$1.380\,658(12) \times 10^{-23} \, \text{J} \, \text{K}^{-1}$
electron specific charge (charge-to-mass ratio)	$-e/m_e$	$-1.758\,819 \times 10^{11} \, \text{C} \, \text{kg}^{-1}$
electron charge (elementary charge)	e	$1.602\,177\,33(49) \times 10^{-19} \, \text{C}$
Faraday constant	F	$9.648\,530\,9(29) \times 10^4 \, \text{C} \, \text{mol}^{-1}$
ice-point temperature	T_{ice}	$273.15 \, \text{K}$ [b]
molar gas constant	R	$8.314\,510(70) \, \text{J} \, \text{K}^{-1} \, \text{mol}^{-1}$
molar volume of ideal gas (at 273.15 K and 101 325 Pa)	V_m	$22.414\,10(19) \times 10^{-3} \, \text{m}^3 \, \text{mol}^{-1}$
Planck constant	h	$6.626\,075\,5(40) \times 10^{-34} \, \text{J} \, \text{s}$
standard atmosphere	atm	$101\,325 \, \text{Pa}$ [b]
speed of light in vacuum	c	$2.997\,924\,58 \times 10^8 \, \text{m} \, \text{s}^{-1}$ [b]

[a] Data are presented in their full precision, although often no more than the first four or five significant digits are used; figures in parentheses represent the standard deviation uncertainty in the least significant digits.

[b] Exactly defined values.

The Periodic Table

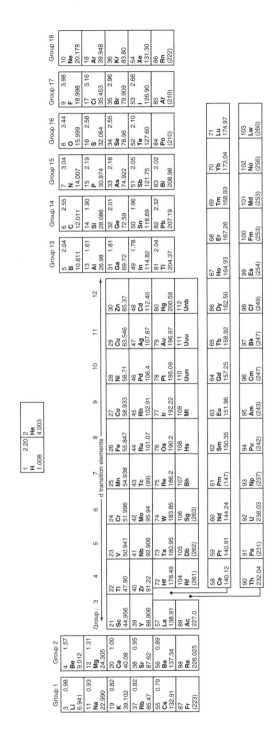

Index